D1200310

COLLOIDAL PHENOMENA

COLLOIDAL PHENOMENA

Advanced Topics

by

C.S. Hirtzel and Raj Rajagopalan

Department of Chemical and Environmental Engineering
Rensselaer Polytechnic Institute
Troy, New York

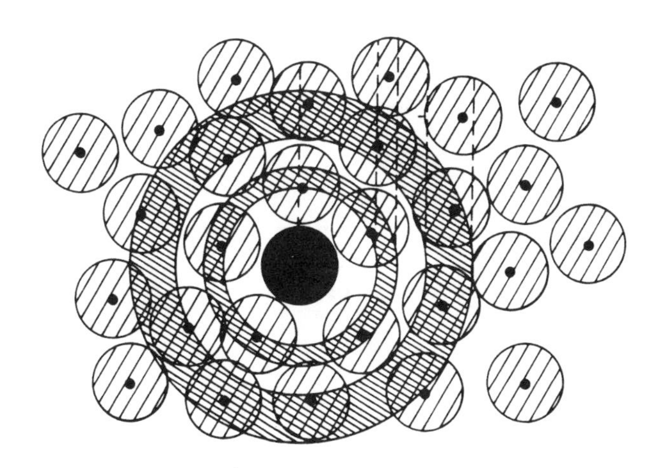

np

NOYES PUBLICATIONS

Park Ridge, New Jersey, U.S.A.

Library of Congress Catalog Card Number: 84-22630
ISBN: 0-8155-1011-X
Printed in the United States

Published in the United States of America by
Noyes Publications
Mill Road, Park Ridge, New Jersey 07656

10 9 8 7 6 5 4 3 2 1

Library of Congress Cataloging in Publication Data

Hirtzel, C.S.
Colloidal phenomena.

Includes bibliographies and index.
1. Colloids. I. Rajagopalan, Raj. II. Title.
QD549.H56 1985 541.3'45 84-22630
ISBN 0-8155-1011-X

To our parents

J.A. Hirtzel and J.M. Hirtzel

R. Nagalakshmi and K. Rajamani

and to

Professor C. Tien

and to the memory of

Professor J.E. Quon

About the Authors

Dr. C.S. Hirtzel is on the faculty of the Department of Chemical and Environmental Engineering at the Rensselaer Polytechnic Institute in Troy, New York. Her areas of research and teaching interests include (in addition to colloidal and interfacial phenomena) probability theory, stochastic processes, and theories of extreme value phenomena and autocorrelated sequences. Dr. Hirtzel is a member of the American Institute of Chemical Engineers, the American Chemical Society, the Society for Industrial and Applied Mathematics, the American Statistical Association and the International Association of Colloid and Interface Scientists and is active in various professional capacities in these professional societies.

Dr. Raj Rajagopalan is also on the faculty of the Department of Chemical and Environmental Engineering at the Rensselaer Polytechnic Institute. His research and teaching activities are in transport phenomena, separation processes, applied mathematics and colloidal and interfacial phenomena. He is a member of the American Chemical Society, the American Physical Society, the International Association of Colloid and Interface Scientists, the Society for Industrial and Applied Mathematics, the American Institute of Chemical Engineers, the Society of Rheology, and the Fine Particle Society and is active in these societies.

PREFACE

The study of interactions in materials of colloidal dimensions has assumed new proportions in recent years for both practical as well as fundamental reasons. With the refinements of numerous experimental tools for probing the static and dynamic structure of dispersions and with the injection of concepts from statistical physics, this field is being recognized and appreciated for what it actually is, namely, a study of structure of matter of macroscopic dimensions. The high degree of interdisciplinary interactions in this endeavor, the potential for cross-disciplinary impact of the results, and the rapid growth of theoretical and experimental advances all point to the need for a consolidation of the ideas and for the identification of promising research directions. It is in view of this that we embarked on the task of preparing a report on some of the above topics for the Office of Interdisciplinary Research of the National Science Foundation about a year ago. The specific scope of this task is outlined in Chapter 1.

The present book is a corrected and slightly revised version of the abovementioned report and was prepared at the request of Noyes Publications. We are pleased that Noyes chose to release this report through normal publishing channels to assure a wider dissemination of the contents than possible otherwise.

During the preparation of the original version of this monograph we received many helpful comments and preprints and reprints of publications from a highly interdisciplinary group of researchers and educators working on various aspects of colloid science. This assistance is acknowledged separately in the following pages. In addition, the support we have received from some of our colleagues during and after the preparation of this report has been a source of constant encouragement to us. In particular, we are especially indebted to Professor Howard Brenner, Willard H. Dow Professor of Chemical Engineering at the Massachusetts Institute of Technology, Professor Egon Matijević of Clarkson University at Potsdam, Professor Chi Tien of Syracuse University, Professor

Frank M. Tiller, M.D. Anderson Professor of Chemical Engineering at the University of Houston, and Professor Raffi M. Turian of the University of Illinois at Chicago. In addition to the encouragement and advice they have given us, they have also taught us a great deal, by instruction and, more importantly, by example. We are grateful for this.

A major reward one expects from being in academia is the excitement of being able to share, if one is fortunate enough, the fruits of one's work with colleagues and students with sympathetic and appreciative ears. We thank our colleague, Professor Michael H. Peters of RPI, for sharing our enthusiasm for this work and for numerous discussions on topics of related interest.

We also would like to thank Janet Paul, Senior Editor at Noyes Publications, for her interest in our work and for her helpful comments and advice during the preparation of this book. Of course, this entire undertaking would not have been possible without the enthusiastic encouragement and support of Dr. Morris S. Ojalvo of the National Science Foundation, and without the financial assistance provided by Mr. Richard Goulet through the Office of Interdisciplinary Research of NSF. We also acknowledge with pleasure the constant support of B. Boa throughout the preparation of this work.

The assistance of Linda Van Buskirk and Michael Alterio in preparing the manuscript in camera-ready form has been invaluable. Finally, this work would not have been possible without the active cooperation and assistance of our spouses.

C.S. Hirtzel
R. Rajagopalan

Department of Chemical and
Environmental Engineering
Rensselaer Polytechnic Institute
Troy, NY 12180-3590

August, 1984

ACKNOWLEDGEMENTS

We would like to acknowledge the financial support provided by the National Science Foundation for the preparation of this Report. In addition, we would like to thank:

Dr. R. Goulet	Office of Interdisciplinary Research National Science Foundation
Dr. M. Ojalvo	Chemical and Process Engineering Division National Science Foundation

for their interest and encouragement throughout this project.

We have been very fortunate in having the skilled and friendly assistance of:

M. Kolb	Department of Chemical and Environmental Engineering, RPI

in putting this Report together. We appreciate, in addition to her skills, her interest and her enthusiasm.

In addition, the excellent illustrations and graphics prepared by:

C. Macklin	Office of Instructional Media, RPI
J. Pond	Office of Instructional Media, RPI
T. Witkowski	Office of Instructional Media, RPI

have contributed greatly to the appearance of this Report.

We would like also to thank our students:

C. Castillo
D. Cates
K.Y. Chan
K. Chari
M. Joyce
J. Uthgenannt
M. Venkatesan

for their assistance and comments.

We wish also to acknowledge the following scientists, who responded to our requests for comments and advice on the contents and preparation of this Report. Many people responded generously with their time, comments, reprints and other materials. Their contributions were invaluable in assembling the material necessary for this Report. We hope that we have represented their opinions and contributions accurately, but we must accept the responsibility for any errors or omissions.

Prof. B.J. Ackerson	Oklahoma State University
Prof. J.L. Anderson	Carnegie-Mellon University
Dr. S.L. Brenner	National Institutes of Health
Dr. S.J. Candau	Université Louis Pasteur, France
Prof. D.Y.C. Chan	Australian National University, Australia
Prof. S-H. Chen	Massachusetts Institute of Technology
Dr. T. Dabros	McGill University, Canada (Visiting)
Dr. E.J. Davis	University of Washington
Dr. E. Dickinson	University of Leeds, U.K.
Prof. S.E. Friberg	University of Missouri–Rolla
Dr. R.D. Gimbel	University of Karlsruhe, F.R.G.
Prof. S. Hachisu	Tsukuba University, Japan
Dr. J.B. Hayter	Institute Laue-Langevin, France
Prof. H. Hoffmann	University of Bayreuth, F.R.G.
Dr. D. Horn	BASF, F.R.G.
Dr. R.J. Hunter	University of Sydney, Australia
Prof. N. Imai	Nagoya University, Japan
Prof. N. Ise	Kyoto University, Japan
Dr. G. Jannink	Laboratoire Leon Brillouin, France
Prof. B. Lindman	University of Lund, Sweden
Prof. J. Lyklema	Agricultural University, The Netherlands
Prof. E. Matijević	Clarkson College of Technology
Dr. A.I. Medalia	Cabot Corporation, Massachusetts
Dr. I. Morrison	Xerox Corporation, Webster, New York
Dr. R.H. Ottewill	University of Bristol, U.K.
Dr. P. Pieranski	Laboratoire de Physique des Solides, France
Dr. P.N. Pusey	Royal Signals and Radar Establishment, U.K.
Prof. W.R. Schowalter	Princeton University
Prof. A. Silberberg	Weizmann Institute, Israel
Prof. I. Sogami	Kyoto Sangyo University, Japan
Dr. E.H.K. Stelzer	Max Planck Institute for Biophysics, F.R.G.
Prof. P. Stenius	Institute for Surface Chemistry, Sweden
Prof. C. Tien	Syracuse University
Prof. F.M. Tiller	University of Houston
Prof. R.M. Turian	University of Illinois at Chicago
Prof. T. Tsang	Howard University
Dr. T.G.M. van de Ven	Pulp and Paper Research Institute of Canada, Canada

Prof. C.J. van Oss	State University of New York at Buffalo
Dr. A. Vrij	van't Hoff Laboratory for Physical and Colloidal Chemistry, The Netherlands
Prof. B.R. Ware	Syracuse University
Prof. D.T. Wasan	Illinois Institute of Technology
Dr. E. Wolfram	Lorand Eotvos University, Hungary
Prof. T.F. Yen	University of Southern California

We thank the following organizations for giving us permission to use some figures and other materials from their publications:

Figure 1.1	Reprinted from Laskowski, J., pp. 315-357 in Surface and Colloid Science, Vol. 12, E. Matijević (Editor), Plenum, New York, 1982. Copyright (1982) Plenum Publishing Corporation.
Figures 2.1 and 2.2	Reprinted from Shah, D.O., pp. 1-12, in Surface Phenomena in Enhanced Oil Recovery, D.O. Shah (Editor), Plenum, New York, 1981. Copyright (1981) Plenum Publishing Corporation.
Figure 3.C.1	Reprinted from Ninham, B.W., J. Phys. Chem. 84, 1423 (1980). Copyright (1980) American Chemical Society.
Figures 3.E.1 and 3.E.2	Reprinted from Sato, T. and Ruch, R., Stabilization of Colloidal Dispersions by Polymer Adsorption, Marcel Dekker, New York, 1980. Copyright (1980) Marcel Dekker, Inc.
Figure 4.B.2	Reprinted from Napper, D.H., pp. 99-128 in Colloidal Dispersions, J.W. Goodwin (Editor), Royal Society of Chemistry, London, UK, 1982. Copyright (1982) The Royal Society of Chemistry.
Figure 4.C.2	Reprinted from Schowalter, W.R., Adv. Colloid Interface Sci. 17, 129 (1982). Copyright (1982) Elsevier Science Publishers, Amsterdam, The Netherlands.
Figures 6.A.1, 6.A.2, and 6.A.4	Reprinted from Pieranski, P., Contemp. Phys. 24, 25 (1983). Copyright (1983) Taylor & Francis, Ltd., London, UK.
Figure 6.A.5	Adapted from Ise, N., pp. 115-132 in Biomimetic Chemistry, Z.I. Yoshida and N. Ise (Editors), Kodansha, Tokyo, Japan, 1983. Copyright (1983) Kodansha Scientific Ltd.

Figure 6.A.9 Reprinted from Ackerson, B.J. and Clark, N.A., Physica 118A, 221 (1983). Copyright (1983) North-Holland Physics Publishing, Amsterdam, The Netherlands.

Figure 6.A.10 Reprinted from Hachisu, S. and Yoshimura, S., Nature 283, 188 (1980). Copyright (1980) Macmillan Journals Ltd., London, UK.

Figure 6.A.11 Adapted from Hachisu, S. and Yoshimura, S., Nature 283, 188 (1980). Copyright (1980) Macmillan Journals Ltd., London, UK.

Figure 6.B.1 Reprinted from Ziman, J.M., Models of Disorder, Cambridge Univ. Pr., New York, 1979. Copyright (1979) Cambridge University Press.

Figures 6.B.2 and 6.B.3 Reprinted from Brown, J.C., Pusey, P.N., Goodwin, J.W. and Ottewill, R.H., J. Phys. A: Math. Gen. 8, 664 (1975). Copyright (1975) The Institute of Physics, London, UK.

Figures 6.D.1 and 7.A.1 Reprinted from Boon, J.P. and Yip, S., Molecular Hydrodynamics, McGraw-Hill, New York, 1980. Copyright (1980) McGraw Hill Publishing Company.

Figure 7.A.2 Reprinted from Pusey, P.N., Phil. Trans. Roy. Soc. London 293A, 429 (1979). Copyright (1979) The Royal Society of London.

Figures 7.B.1 and 7.B.2 Reprinted from Reed, C.C. and Anderson, J.L., AIChE J. 26, 816 (1980). Copyright (1980) The American Institute of Chemical Engineers, New York.

Figure 7.C.1 Reprinted from Goodwin, J.W., pp. 165–195 in Colloidal Dispersions, J.W. Goodwin (Editor), Royal Society of Chemistry, London, UK, 1982. Copyright (1982) The Royal Society of Chemistry.

Figures 7.C.2 and 7.C.3 Reprinted from Mewis, J. and Spaull, A.J.B., Adv. Colloid Interface Sci. 6, 173 (1976). Copyright (1976) Elsevier Science Publishers, Amsterdam, The Netherlands.

Figure 7.C.5 Reprinted from Russel, W.B., J. Rheol. 24, 287 (1980). Copyright (1980) The Society of Rheology, Inc., NY.

Table 7.C.1 Reprinted from Hoffman, R.L., pp. 578-588 in G.W. Poehlein, R.H. Ottewill and J.W. Goodwin (Editors), Science and Technology of Polymer Colloids: Characterization, Stabilization and Application Properties, Vol. II, Martinus Nijhoff, The Hague, The Netherlands, 1983. Copyright (1983) Martinus Nijhoff Publishers.

Contents and Subject Index

Part II
Dilute Dispersions

Part III
Concentrated Dispersions

SCOPE

1

Scope

CHAPTER 1

SCOPE

This report presents a review of research activities in the United States and abroad in a few selected areas of colloidal interactions and phenomena. The problems of scientific and industrial relevance in which properties and behavior of species of colloidal size play a dominant role are numerous, each with its own subtleties and special requirements. An example, taken from an area of considerable current activity, namely, physicochemical separation processes, is presented in Figure 1.1 and illustrates the assortments of individual problems, within a single general area, that require an understanding of colloidal interactions. A review of all such areas of application is beyond the scope of this Report. In order to keep this review and its usefulness tractable, the scope of this review will be restricted to the following technical aspects:

- Colloidal phenomena relevant in adsorption, deposition and reentrainment of species of colloidal dimensions.
- Properties of dispersions at finite concentrations, since most dispersions of interest in industry and nature are very seldom dilute. Even when they are 'dilute', their properties are very often controlled by complex many-body forces, which make the dispersions effectively 'concentrated'.
- Experimental and theoretical tools that are of particular relevance to the above aspects, from the point of view of both laboratory-scale research investigations and larger-scale practical extensions.

The above focus, despite its restrictions, covers a major portion of the technical areas in research activities that are currently in progress in the United States and abroad (see Table 1.1). Within this general focus, the specific objectives of this Report are to provide:

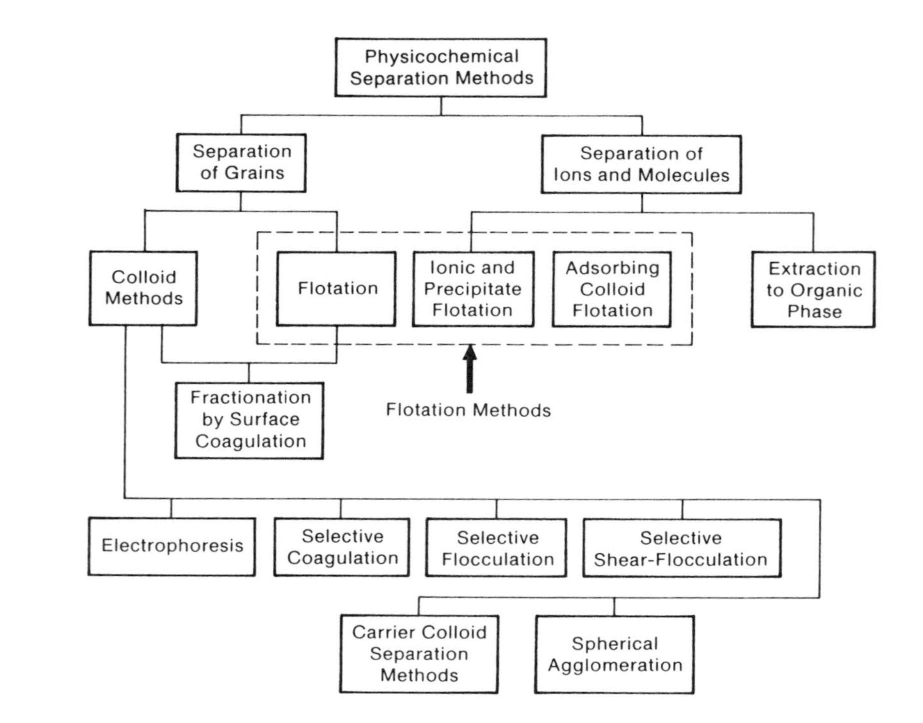

Figure 1.1 An example of the impact of colloidal phenomena in physical and chemical separation processes.

[From Laskowski, J., pp. 315-357 in Surface and Colloid Science, Vol. 12, Matijević, E., (ed.), Plenum, New York, NY, 1982.]

TABLE 1.1

COLLOIDAL PHENOMENA IN SCIENCE AND ENGINEERING

General Areas	Some Current Problems of Interest	Disciplinary Label
Separation processes	Particulate deposition; colloidal stability; viral and bacterial adhesion	(PROCESS) ENGINEERING; BIOLOGY
Fermentation processes	Adsorption of microorganisms; protein processing; multiphase processing in biochemical reactors	(BIOCHEMICAL and REACTION) ENGINEERING
Membrane processes	Fouling of membranes; charge effects on separation efficiency	(ENVIRONMENTAL) ENGINEERING; COLLOID CHEMISTRY
Concentrated dispersions	Steric stability; transport properties; Rheological behavior; order/disorder transformations	FLUID MECHANICS; PHYSICS; COLLOID CHEMISTRY
Analytical techniques and fundamental aspects	Particle size analysis; Dynamic light scattering and other scattering techniques; Electrophoresis; Structure and dynamics of ionic double layers	CHEMISTRY; PHYSICS

(i) a review of current activities in fundamental research relevant to the above areas of focus,

(ii) an identification of interdisciplinary and cross-disciplinary research needs and long-term research directions, and

(iii) a source of information transfer between researchers in areas of application and those in basic, support sciences.

The interdisciplinary nature of research and research needs in the areas of interest herein is indicated in Table 1.1 and will be evident from the material presented in the subsequent Chapters of this Report. For instance, the study of concentration effects in dispersions is, in a fundamental sense, a study of the structure of matter of colloidal dimensions. Thus, it is not surprising that the rapid growth of research in this area in recent years is largely due to the interdisciplinary interaction between chemists and physicists with a background in liquid-state physics and scientists and engineers with interests in colloidal phenomena. The transfer of information and know-how, however, has not been one-sided, for the rewards of this cooperation have had a beneficial consequence for both groups. For instance, this interdisciplinary interaction not only has been a catalyst in the rapid growth of understanding of many-body effects in colloidal systems (e.g., application of theories of liquid-state physics in the study of colloidal interactions) but has also led to the development of promising new tools for the study of molecular systems (e.g., use of colloidal crystals as model many-body systems). Specific examples of this information exchange can be found in Chapters 6 and 7.

The rest of this Report is divided into three Parts, each containing two Chapters, and a final concluding Chapter that presents a summary of major recommendations and a brief list of some important problems that fall outside the scope of this review. Part I on Colloidal Interactions presents, in Chapter 2, a brief summary of typical colloids and colloidal phenomena and a list of major textbooks and research monographs in this area. Chapter 3 reviews the current status of research on intermolecular, electrokinetic, steric and other forces needed for the material discussed in the other Chapters, and, in addition, serves as a common link for the rest of the Report. Parts II and III focus on dilute and concentrated dispersions, respectively. Colloidal stability, which has been traditionally the central focus of colloid science, is the subject of Chapter 4. The related problem of deposition and resuspension of colloidal species on substrates follows in Chapter 5. The discussions on concentrated dispersions, presented in Part III, have been divided, somewhat artificially, into two Chapters. The first focuses on many-body theories and experimental studies of the static and dynamic structure of dispersions. Some selected transport properties are discussed in the other.

The above organization permits each Chapter to focus on a major area of current activity while at the same time being grouped with other problems of related interest. Each Chapter is technically self-contained, has its own list of references (grouped separately under each major subdivision of the Chapter), presents its own recommendations and can be read without reference to the other Chapters. However, numerous cross-references to other Chapters are made throughout each Chapter to point out similarities and problems of mutual interest and to establish the needed or potential links. The general approach in each Chapter (with the exception of Chapter 2, which is introductory in nature) has been to identify:

- the major subdivisions of the area of activity addressed in the

Chapter,

- the major components of each subdivision,
- the major technical problems in each component, and
- the key references and the major accomplishments on each technical problem.

A list of research needs is given at the end of each Chapter.

Our goal throughout this Report has been to present the material at a level that would satisfy the specialist and yet, at the same time, would permit a relative newcomer to the area to follow the rationale for the ongoing research in the area and what additional work needs to be done. We have avoided presenting equations (e.g., expressions for double layer interactions) and tables of data since it is impossible to present a balanced account of these in the limited space available here. However, numerous illustrations, many of which are based on actual theoretical and experimental results, are presented throughout the Report to emphasize some key points. The purpose of these is to illustrate the qualitative aspects of the phenomena and concepts discussed; detailed quantitative results can be found in the references cited.

Despite its limited scope, the Report addresses a large part of colloid science; yet, many problems of substantial importance (such as thin film phenomena, foam stability, etc.) have been excluded. We hope that this Report will be a first step in the preparation of a comprehensive review of all facets of colloid science. We are also aware that we may have missed some important references and research activities that are relevant to the material covered in this Report. We would appreciate hearing from readers who come across such omissions (or inaccuracies) so that these can be corrected in the future editions.

PART I

COLLOIDAL INTERACTIONS

2

Colloids and Colloidal Phenomena

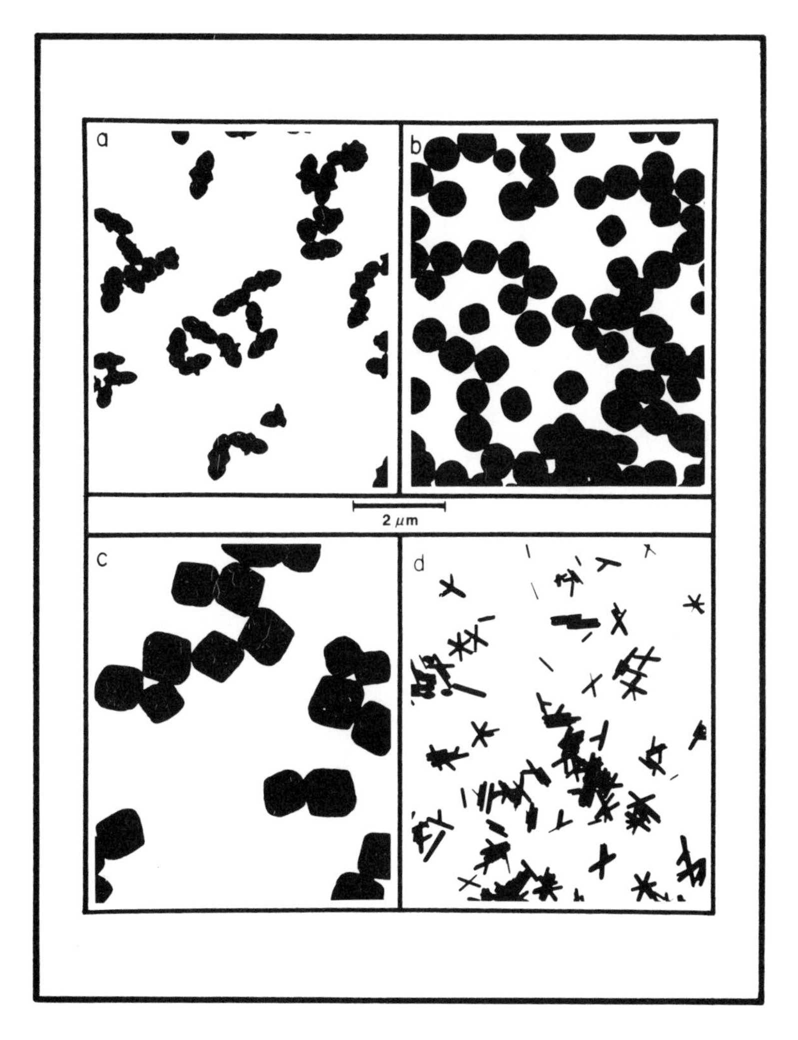

CHAPTER 2

COLLOIDS AND COLLOIDAL PHENOMENA

Overview

The subject matter of this Report is, in very broad terms, colloids, i.e., colloidal interactions and phenomena. Colloidal systems include systems containing very small particles or fibers, or large or macromolecules, or thin films. More specifically, a colloid, as defined by the International Union of Pure and Applied Chemistry, is any material for which one or more of its three dimensions lies within the range of 1 to 1000 nanometers. Obviously, the array of subjects and problems which fall within the general label of colloids or colloid science is vast, and it would be difficult for a single report or monograph to treat all of these in a comprehensive and adequate manner. The scope of this Report, as explained in Chapter 1, is thus restricted to a study of research and research needs in a certain class of colloidal interactions. The focus of this Report is on the interaction forces between species of colloidal dimensions and how these interactions affect stability, deposition, order/disorder transitions, and transport properties. There are numerous other problems which require attention, but these fall outside the scope of this Report [e.g., thin film phenomena such as foams, and the thermodynamics and formation of micelles and microemulsions]. We hope that this Report will serve as a first step in a comprehensive review of many of these other aspects of colloidal and interfacial phenomena that merit attention.

The intent of this present Chapter is to present a very brief review of relevant background material, including examples of colloidal systems, examples of processes in which colloidal interactions and phenomena are significant, classification of colloidal systems, and other background information on the colloidal state.

The study of colloidal interactions and phenomena is very much interdisciplinary in nature, both from the point-of-view of the underlying fundamental principles involved in understanding colloidal interactions, as well as from the point-of-view of the applications in which these interactions are significant. Research in colloidal phenomena draws upon many fields of science including chemistry, physics, biology, geology, and mathematics, and diverse fields of engineering science and applications. Numerous examples of these are discussed throughout this Report. To illustrate this interdisciplinary nature and the wide range of processes where colloidal phenomena are significant, selected examples of colloidal systems are given in Table 2.1. Moreover, some examples of technological and other processes in which applications are important are presented in Table 2.2. These lists are by no means complete and are intended only to give an idea of the numerous, diverse areas of science and technology in which colloidal systems are encountered, and of the myriad applications in which colloidal interactions play a role.

As well-known, these phenomena are highly complex and depend on many physical and chemical (and, in some cases, biological) factors. The properties which are of primary significance in determining their nature and behavior include particle size, particle shape, the flexibility (or lack of) of the particle, the chemical and electrical properties of the particle surface (such as charge and distribution of charge), the interactions of the particle with other particles, and the interactions of the particle with the solvent, i.e., the medium in which the particles are dispersed. Other relevant factors include the distribution of particle size in the system (e.g., the degree of

TABLE 2.1

SELECTED EXAMPLES OF COLLOIDAL SYSTEMS

Type of System	Dispersed Phase	Medium	Examples
Aerosols:			
Liquid Aerosol	Liquid	Gas	Fogs, liquid sprays, mists
Solid Aerosol	Solid	Gas	Dusts, smokes
Foams:			
Gas Foam	Gas	Liquid	Fire-extinguishing foams
Solid Foam	Gas	Solid	Expanded polystyrene
Emulsions:			
Liquid Emulsion	Liquid	Liquid	Milk and dairy products, mayonnaise, oil in water, water in oil
Solid Emulsion	Liquid	Solid	Opals, pearls
Suspensions:			
Colloidal Suspension	Solid	Liquid	Cement, clays, cosmetic and pharmaceutical preparations, dyes, ink, paints
Solid Suspension	Solid	Solid	Plastics, alloys
Association Colloids	Surfactant assemblies	Liquid	Micelles, inverse micelles

Note: Numerous other colloidal systems and examples such as thin films, soap bubbles, biological membranes, and others have not been included for the sake of brevity.

TABLE 2.2

SELECTED EXAMPLES OF PROCESSES IN WHICH COLLOIDAL PHENOMENA ARE SIGNIFICANT
Adhesion
Adsorption processes
Ceramics
Chromatography
Delta formation
Detergency
Emulsion polymerization
Food processing
Fouling in heat exchangers
Grinding
Heterogeneous catalysis
Ion exchange
Liquid crystals
Lubrication
Membrane phenomena
Oil recovery
Oil-well drilling
Ore flotation
Powder metallury
Precipitation
Separation processes
Soils conditioning
Sugar refining
Transport of dispersions
Water and waste-waters treatment
Wetting

polydispersity of the system), among others. Colloidal systems also exhibit various other properties of interest including optical properties (such as turbidity or color), kinetic properties (such as Brownian motion or translational and rotational diffusion), electrokinetic properties (related to the motion of the particle in an electrical field), and rheological or mechanical properties (such as viscosity of the dispersion).

Classification of Colloidal Systems

The most general scheme for classifying colloidal systems is to divide these into two categories: (1) lyophilic (or, if the dispersion medium is water, hydrophilic); and (2) lyophobic (or, if water is the dispersion medium, hydrophobic). The lyophilic class of colloidal systems includes several types of colloids, for example:

(i) the simplest type is that of a colloidal system which is formed by a suspension of macromolecules; this includes dispersions of polymers or biopolymers such as proteins; and

(ii) solutions of soaps and/or other amphipolar materials (these are molecules which are characterized by a large non-polar portion and a polar part which is compatible with the dispersion medium, e.g., water). These are referred to as association colloids (or sometimes as colloidal electrolytes). These include micelles (i.e., aggregates of approximately twenty or more molecules in a medium of water) and inverse micelles (for substances in a non-polar medium). Microemulsions also fall within this category (see Figures 2.1 and 2.2).

The second class of colloidal systems, i.e., lyophobic colloids, includes, for example, the gold sols originally prepared and studied by Faraday and other early colloid scientists. Lyophobic colloids are prepared by dispersion or by condensation methods (see Figure 2.3; also see the General References listed at the end of this Chapter).

The above discussion will suffice to serve as the background for the

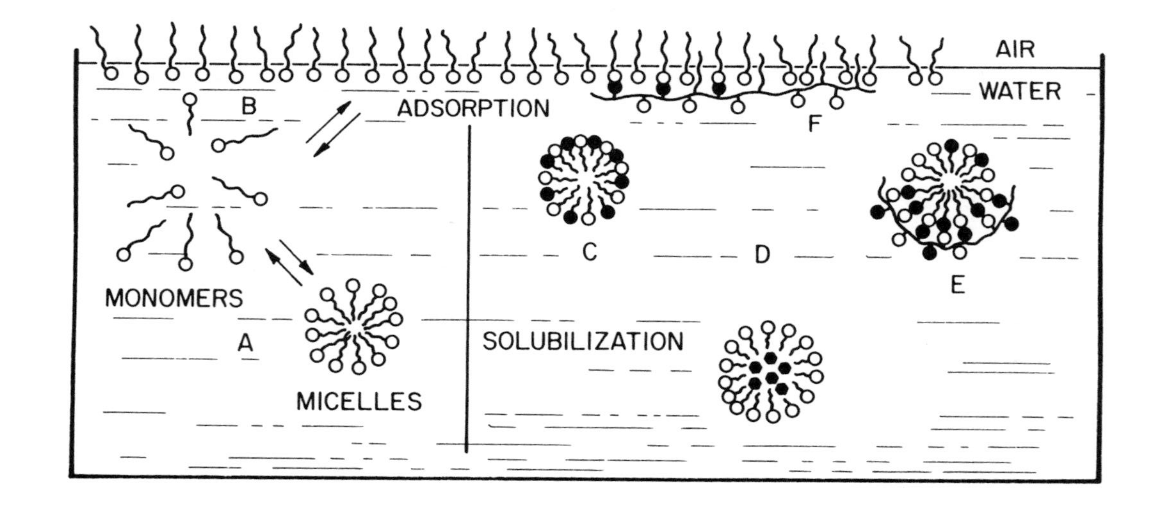

Figure 2.1 Formation of association colloids [see Shah (1981)].

Formation of micelles (A) at large concentrations of the surfactants is illustrated schematically. At low concentrations, the monomers in the bulk are at equilibrium (B) with the molecules adsorbed at the gas/liquid interface. Solubilization of oils in the micelle (D) and adsorption of long-chain polymers (E) and other possibilities are also shown.

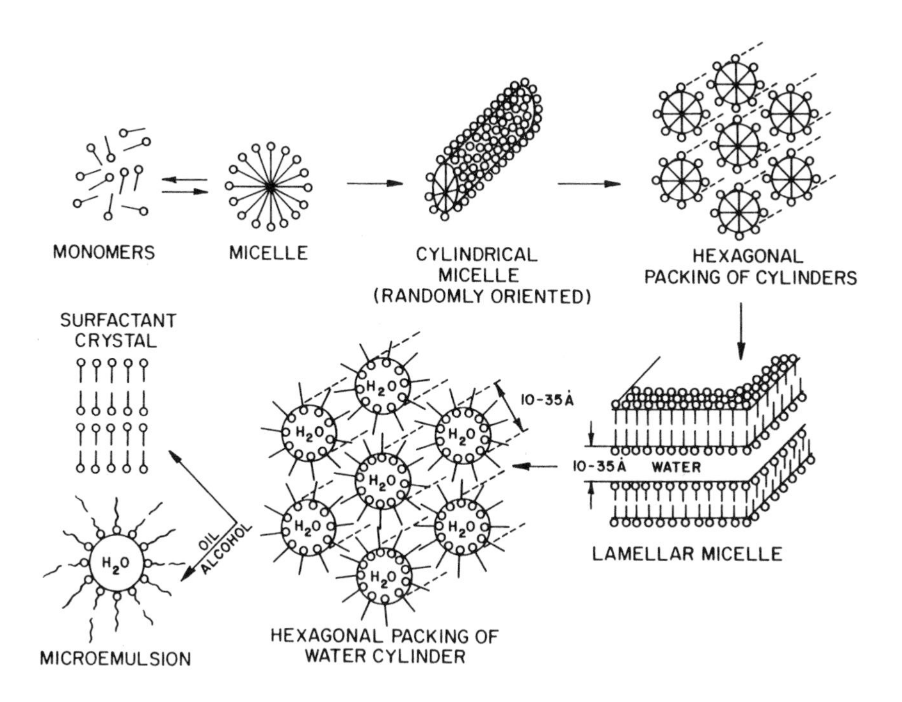

Figure 2.2 Structural transitions in association colloids [see Shah (1981)].

As the concentration of the surfactants is increased, several structural transitions in the assemblies occur. Some of these are illustrated above.

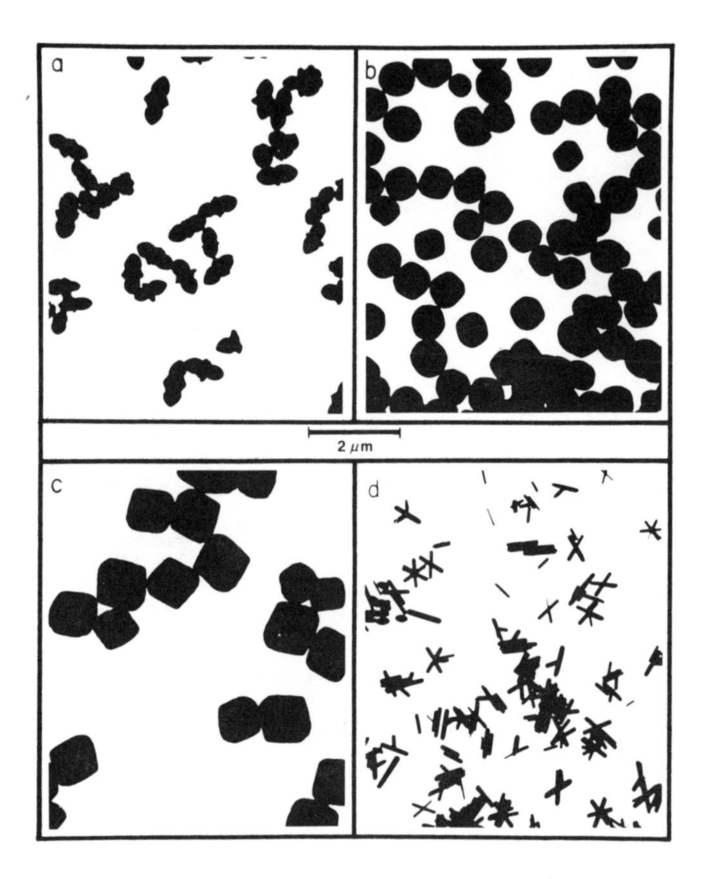

Figure 2.3 Electron micrographs of particles obtained by aging solutions of $FeCl_3$ and HCl under various conditions.

Photographs courtesy of Professor Egon Matijević of Clarkson University, Potsdam, New York. [See Matijević, E., J. Colloid Interface Sci. 58, 374 (1977) for details.]

remaining chapters of this Report. However, before proceeding further, a few remarks are made with respect to the usage of the term 'stability' with respect to colloidal systems. The term stability can be used in two ways; i.e., from the point-of-view of thermodynamic stability, or secondly, from the point-of-view of the kinetic stability of the system. This report focuses primarily on lyophobic systems, although the statistical mechanical theories of interacting dispersions discussed in Chapter 6 do apply to microemulsions and micellar solutions and can, in fact, be used to study the intra-micellar structure, aggregation number, and the like.

Finally, before proceeding to subsequent chapters, a list of General References is presented in the next Section of this Chapter. This list includes textbooks as well as monographs and books at a research level.

General References

Berkeley, R. C. W., Lynch, J. M., Melling, J., Rutter, P. and Vincent, B. (eds.), Microbial Adhesion to Surfaces, Wiley, New York, NY, 1981.

Bitton, G. and Marshall, K. C. (eds.), Adsorption of Microorganisms to Surfaces, Wiley-Interscience, New York, NY, 1980.

Cadenhead, D. A. and Danielli, J. F. (eds.), Progress in Surface and Membrane Science, Vols. 10-14, Academic Press, New York, NY, 1976-1981.

Danielli, J. F., Pankhurst, K. G. A. and Riddiford, A. C. (eds.), Recent Progress in Surface Science, Vols. 1 and 2, Academic Press, New York, NY, 1964.

Danielli, J. F., Riddiford, A. C. and Rosenberg, M. D. (eds.), Progress in Surface and Membrane Science, Vol. 3, Academic Press, New York, NY, 1970.

Danielli, J. F., Rosenberg, M. D. and Cadenhead, D. A. (eds.), Progress in Surface and Membrane Science, Vols. 4-9, Academic Press, New York, NY, 1971-1975.

Dickinson, E. and Stainsby, G., Colloids in Food, Applied Science Publishers, London, U. K., 1982.

Ellwood, D. C., Melling, J. and Rutter, P. (eds.), Adhesion of Microorganisms to Surfaces, Academic Press, London, U. K., 1979.

Everett, D. H. (ed.), Specialist Periodical Reports, Colloid Science, Vols. 1, 2 and 3, The Chem. Soc., London, U. K., 1973, 1975 and 1979.

Everett, D. H. (ed.), Specialist Periodical Reports, Colloid Science, Vol. 4, The Royal Soc. of Chem., London, U. K., 1983.

Fitch, R. M. (ed.), Polymer Colloids, Plenum, New York, NY, 1971.

Fitch, R. M. (ed.), Polymer Colloids II, Plenum, New York, NY, 1980.

Good, R. J. and Stromberg, R. R. (eds.), Surface and Colloid Science, Vol. 11, Experimental Methods, Plenum, New York, NY, 1979.

Goodwin, J. W. (ed.), Colloidal Dispersions, The Royal Soc. of Chem., London, U. K., 1982.

Hiemenz, P. C., Principles of Colloid and Surface Chemistry, Marcel Dekker, New York, NY, 1977.

Hunter, R. J., Zeta Potential in Colloid Science, Academic Press, London, U. K., 1981.

Kerker, M. (ed.), Surface Chemistry and Colloids, Butterworths, London, U. K., 1972.

Kerker, M. (ed.), Colloid and Interface Science, Vols. 2, 3, 4 and 5, Academic Press, New York, NY, 1976.

Kerker, M., Zettlemoyer, A. C. and Rowell, R. L. (eds.), Colloid and Interface Science, Vol. 1, Academic Press, New York, NY, 1977.

Kruyt, H. R. (ed.), Colloid Science, Vol. 1: Irreversible Systems, Elsevier, Amsterdam, The Netherlands, 1952.

Kruyt, H. R. (ed.), Colloid Science, Vol. 2: Reversible Systems, Elsevier, Amsterdam, The Netherlands, 1949.

Mahanty, J. and Ninham, B. W., Dispersion Forces, Academic Press, London, U. K., 1976.

Matijević, E., (ed.), Surface and Colloid Science, Vols. 1-9, Wiley-Interscience, New York, NY, 1969-1976.

Matijević, E., (ed.), Surface and Colloid Science, Vol. 10, Plenum, New York, NY, 1978.

Matijević, E., (ed.), Surface and Colloid Science, Vol. 12, Plenum, New York, NY, 1982.

Ninham, B. W., Overbeek, J. Th. G. and Zettlemoyer, A. C. (eds.), Interaction of Particles in Colloidal Dispersions, Parts A and B, Adv. Colloid Interface Sci. (Special Issues) 16, 17, 1982.

Poehlein, G. W., Ottewill, R. H. and Goodwin, J. W. (eds.), Science and Technology of Polymer Colloids, Vols. I and II, Martinus Nijhoff, The Hague, The Netherlands, 1983.

Sato, T. and Ruch, R., Stabilization of Colloidal Dispersions by Polymer Adsorption, Marcel Dekker, New York, NY, 1980.

Shah, D. O. (ed.), Surface Phenomena in Enhanced Oil Recovery, Plenum, New York, NY, 1981.

Shaw, D. J., Introduction to Colloid and Surface Chemistry, 3rd ed., Butterworths, London, U. K., 1980.

Sonntag, H. and Strenge, K., Coagulation and Stability of Disperse Systems, Halstead Press, New York, NY, 1972.

The Faraday Division, Chemical Society, Colloid Stability, Faraday Discuss. Chem. Soc. No. 65, London, U. K., 1978.

van Olphen, H. and Mysels, K. J. (eds.), Physical Chemistry: Enriching Topics from Colloid and Surface Science, IUPAC Commission I.6, Theorex, La Jolla, CA, 1975.

Verwey, E. J. W. and Overbeek, J. Th. G., Theory of the Stability of Lyophobic Colloids, Elsevier, Amsterdam, The Netherlands, 1948.

Vold, R. D. and Vold, M. J., Colloid and Interface Chemistry, Addison-Wesley, Reading, Mass., 1983.

3

INTERMOLECULAR AND ELECTROKINETIC FORCES

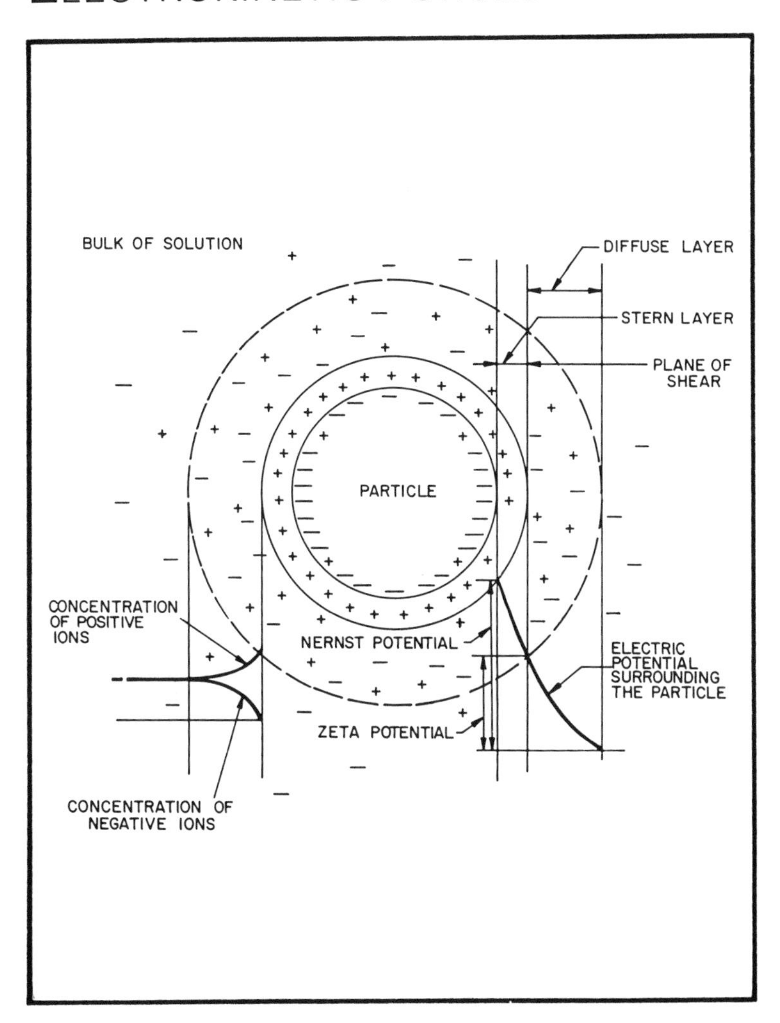

CHAPTER 3

INTERMOLECULAR AND ELECTROKINETIC FORCES

Overview

Investigations into the nature of the forces between colloidal particles have a primary role in many areas of physical, chemical, and engineering science, including colloid science. Understanding and description of these forces are essential elements in the evaluation of stability of a colloidal suspension or of other properties of the dispersion such as its internal structure, order, rheological properties, and others. In the 1940's, for example, the development of the DLVO theory [see Chapter 4 of this report] for the description of colloidal stability focused attention on the role of surface forces and the need for their evaluation. More recently, interest has been focused on short-ranged forces, in addition to Born repulsion, due to solvent structure (termed 'structural forces' herein) and their measurement and role in repulsive interactions.

The subject of this Chapter is a discussion of the various intermolecular and electrokinetic forces which are relevant to colloidal interactions and phenomena. These forces are classified into five categories herein; these are:

(A) Electric Double Layer Structure and Forces,

(B) Attractive Surface Forces,

(C) Structural Forces,

(D) Forces Relevant to Biological Systems, and

(E) Steric Forces.

A. Electric Double Layer Structure and Forces

The electric double layer and its properties play a primary role in many colloidal phenomena and are essential in interpreting colloid stability and the electrokinetic properties of charged colloidal dispersions. The formation and structure of the electric double layer at a charged particle or interface may be qualitatively and briefly described as follows. The colloidal particle has some surface electric charge associated with it. This surface electric charge exerts an influence on the distribution of charge (ions) in the solution, or medium. An electric double layer then forms due to coulombic forces (i.e., the attractive and repulsive forces of unlike and like charges) and entropic forces, the latter of which result in a tendency for the ions to mix and to spread uniformly throughout the available space. Other forces may also be involved in the formation of the electric double layer at a charged interface. Lyklema (1982) terms all forces which are neither coulombic nor entropic in nature as 'specific'. Such specific forces may be either chemical or physical in nature.

In this section, we present, first, a brief discussion of selected phenomenological aspects of electric double layers. Second, a brief overview of theoretical models for the distribution of charge in the electric double layer, and for the magnitude of the electric potentials near the charged interface, is given. Third, theoretical and experimental results relevant to the study of electrokinetic phenomena such as electrophoresis, and the role of the electric double layer in these phenomena, are presented.

For additional material on electric double layers, the reader is referred to the recent review (on relaxed, isolated electric double layers) by Lyklema (1982), and references cited therein. Other relevant reviews are given in Bockris, Conway and Yeager (1980). Additional material on structure and properties of double layers is available in the monograph by Hunter (1981) and in texts such as Hiemenz (1977), Shaw (1980) or Vold and Vold (1983), or in classics such as Overbeek (1952) or Kruyt (1952).

A.1 Phenomenological Aspects of Electric Double Layers

As noted above, the electric double layer results when a (particle) surface or interface acquires a charge. There are several mechanisms through which surfaces may acquire a charge. These include preferential adsorption of ions, dissociation of surface groups, isomorphic substitution of ions, accumulation (or depletion) of electrons at the interface, and adsorption of polyelectrolytes or charged macromolecular species. Other related mechanisms may also effect the nature of the electric double layer. For example, adsorption of molecules which are dipolar may affect the distribution of charge, if not the net surface charge.

As well-known, the electric double layers are characterized by the parameters of charge and potential. A third parameter, the capacitance, is also frequently used as a measure of the degree of the screening of the surface charge by the countercharge. Other properties of, and parameters used to describe, electrical double layers include the point of zero charge, isoelectric points, among others. Properties of the electric double layer, their interrelationships and experimental results of electrokinetic measurements are reviewed in the article by Lyklema cited above, and in the books and monograph cited earlier.

A.2 Models for Distribution of Charge in Electric Double Layer

The structure of the electric double layer is, as well-known, generally regarded as consisting of two regions: an outer, diffuse layer known as the Gouy (or Gouy-Chapman) layer, and an inner, compact layer known as the Stern layer (see Figure 3.A.1). The distributions of ions in these two sections of the double layers are governed by complex physical and chemical variables and their interactions. Usually, however, it is assumed that since the chemical forces are relatively short-ranged in nature, they are significant only in the layer next to the surface (i.e., the Stern layer). On the other hand, the distribution of ions in the outer, diffuse layer is usually treated according to an ideal model. This outer, ideal layer begins at some distance, d, from the surface. The model of the diffuse layer proposed by Gouy (1910) and Chapman (1913) is based on five assumptions as follows: (1) the surface is an infinite, flat plate and is uniformly charged; (2) the solvent exerts influence on the double layer through its dielectric constant (only); (3) the value of the dielectric constant is constant throughout the diffuse layer; (4) ions in the diffuse layer are point charges with a distribution described by the Boltzmann distribution; and (5) the electrolyte is assumed to be a single symmetrical electrolyte. Using the Boltzmann distribution to describe ionic concentration, and Poisson's equation to relate charge density and electric potential, expressions which describe the decay of potential with distance from the charged surface are derived. Similarly, expressions which relate surface potential and surface charge density to distance, ionic concentration and other parameters are developed [see, for example, Overbeek (1952); Shaw (1980); Lyklema (1982); or Hunter (1981)].

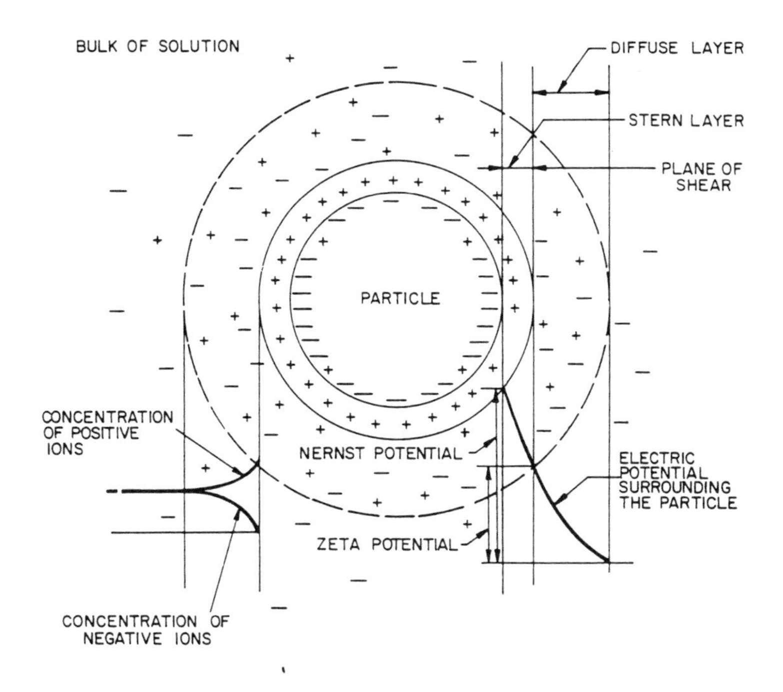

Figure 3.A.1 A schematic representation of the electrical double layer around a particle.

Combination of the Poisson equation for electrostatic potential and the Boltzmann equation for charge density yields the well-known Poisson-Boltzmann equation (PBE) for the electrostatic potential [e.g., Verwey and Overbeek (1948); Shaw (1980); Hunter (1981)].

The general PBE above has no analytical solution. However, for the case where the potential is low (i.e., less than approximately 25 mV at 25°C), the Poisson-Boltzmann equation can be linearized through a power series expansion of the exponential term (i.e., this is the Debye-Hückel approximation). In this case, the resulting solution of the PBE yields the result that the potential decreases exponentially with distance from the charged surface. However, for very small distances from the charged surface, the potential is relatively high and the Debye-Hückel approximation above does not apply; and the rate of decay of electrostatic potential with distance from the charged surface is greater than exponential.

The results described above are for a <u>flat</u> double layer. Expressions for the solution of the Poisson-Boltzmann equation for a spherical interface are given, for example, in Vold and Vold (1983), Shaw (1980), and others. Again, for the case where the potential is low, the Debye-Hückel approximation can be used and the resulting expressions can be solved analytically. However, the Debye-Hückel approximation frequently does not hold for cases of interest to the colloid scientist; Loeb, Overbeek and Wiersema (1961) have obtained <u>un</u>approximated solutions to the Poisson-Boltzmann equation for spherical interfaces.

Stern Layer

For the inner, compact portion of the electric double layer, the treatment is complex. As stated above, the development of the Gouy-Chapman theory of the diffuse layer assumes point charges in the electrolyte medium. In fact, however, ions have finite sizes; this constraint will affect the distance from the surface of the inner boundary of the diffuse double layer. Stern (1924) developed a model such that the electric double layer consists of two parts; the interface between the layers, the Stern plane, is located at a distance of approximately one hydration radius away from the surface and separates the two layers. In the outer layer, the theory of Gouy-Chapman applies. However, the treatment of the inner layer is more complicated. Different cases are possible; e.g., no specific ion adsorption occurs, some specific ion adsorption occurs, or superequivalent specific ion adsorption occurs, to name a few [see, for additional details, Lyklema (1982) and references therein]. Expressions for the Stern model of the double layer are in texts such as Shaw (1980). The Stern model provides a useful, if rough, model for the interpretation of the inner layer and experimental observations of double layer phenomena. One important refinement to the Stern model is that of Grahame (1947). Grahame distinguishes, within the double layer, an 'inner Helmholtz plane' and an 'outer Helmholtz plane'. The former is representative of the plane through the centers of specifically adsorbed ions which are usually dehydrated upon adsorption; the 'outer Helmholtz plane' is the plane parallel to the surface (at distance d) and corresponds to the distance of closest approach of the hydrated ions (i.e., the Stern plane in Stern's terminology).

Modified PBE and Monte Carlo Studies of Double Layers

The assumption that the potential in the diffuse part of the double layer can be reasonably well-represented by the solution of the Poisson-Boltzmann equation (as suggested by Gouy and Chapman) is generally believed to be true below electrolyte concentrations of the order of 10^{-2}M for 1-1 electrolytes [Napper and Hunter (1972)]. The initial study of Krylov and Levich (1963) for concentrations in the range of 0.1-0.5M seems to indicate that the potential drop is sharper at these concentrations than would be predicted by the Gouy-Chapman theory. Other attempts in this direction have been made, but Levine and Bell (1966) emphasize that one must account for all relevant effects, such as volume of ions, variation of permittivity, ion self-atmosphere effects and electrostriction, since these tend to counterbalance each other. It is estimated currently that the simple Gouy-Chapman theory is sufficient to describe the potential outside the outer Helmholtz plane for 1-1 electrolytes. For higher valencies, the corrections may be large and uncertain, and statistical mechanical computations based on Monte Carlo methods are now being used to assess the various double layer theories, both old (cited above) and new [see Outhwaite and Bhuiyan (1983)]. Some of the recent double layer theories are discussed in Levine and Outhwaite (1978), Outhwaite and Bhuiyan (1982), Carnie et al. (1981), Henderson and Blum (1978, 1981), Levine, Outhwaite and Bhuiyan (1981), and Croxton and McQuarrie (1979, 1981). The results based on Monte Carlo simulations may be found in Torrie, Valleau and Patey (1982), Torrie and Valleau (1982), Croxton et al. (1981), van Megen and Snook (1980) and Snook and van Megen (1981). Good agreement between the recent, modified versions of the Poisson-Boltzmann theories and the computer simulations for 1-1 electrolytes has been reported, but discrepancies persist in the case of higher-valency electrolytes. The recent work of Outhwaite and Bhuiyan (1983) provides an additional approximate solution that leaves the structure of the modified Poisson-Boltzmann equation unchanged but modifies its value near the surface. This improves the previous solutions and predicts the diffuse layer potential drop reasonably well for unsymmetrical electrolytes for low values of surface charges; however, discrepancies remain at higher values of surface charges. Numerical results for mean electrostatic potentials under various conditions of valency and surface charge are presented in the above paper, but more work is needed in the case of unsymmetrical electrolytes.

A.3 Interacting Double Layers

In a dispersion, as particles approach each other, they begin to influence each other as soon as the double layers overlap. That is, repulsive electrostatic forces arise as charged particles move closely together due to interactions of the surface charges of the particles and the diffuse portions of the electric double layers. [The inner, or Stern, layer does not have a direct role in the interactions between surfaces since it is only several molecules thick; however, it has a significant indirect role in these interactions in determining the value of the potential of the diffuse portion of the double layer. For additional details, see Lyklema (1982).] The Poisson-Boltzmann equation for the electrostatic potential in the diffuse layer was discussed above. For two interacting surfaces, the electrostatic potential energy at any position x is the change in the free energy of the system as the two double layers are brought to a distance of x apart from infinite separation.

The magnitude and the shape of the interaction energy profile which results from the analysis also depend on the boundary conditions which are assumed to describe the interacting surfaces. For example, interactions may occur for (1) surfaces which maintain constant potentials as they approach each other; (2) surfaces which maintain constant charge densities as they approach; or (3) mixed types of interactions. Approximate expressions for interaction at constant potential (i.e., case 1) between two parallel plates and between two dissimilar spheres were first presented by Hogg, Healy and Fuerstenau (1966). The case of interaction at constant charge (i.e., case 2) was considered by Wiese and Healy (1970); they extended the results of Hogg, Healy and Fuerstenau to the case of constant charge interaction using a linearized version of the Poisson-Boltzmann equation and results of Frens (1968). Other expressions for interactions at constant charge are available in Honig and Mul (1971), Bell and Peterson (1972), and in Gregory (1975). The mixed interaction mode (i.e., case 3), in which one surface maintains constant potential while the other maintains constant charge density during approach, has been considered by Kar, Chander and Mika (1973). The question as to which of these various modes of interaction actually occurs is an important area of current colloid research, since it has a direct impact on colloid stability (see Chapter 4, Section A) and deposition phenomena (see Chapter 5 and Figure 5.2).

Other phenomena may also be important for understanding the electrostatic double layer interactions. For example, Lyklema (1982) cites experimental results on silver iodide solutions which seem to indicate that intermediate situations, in which neither the surface potential nor the surface charge remains constant during interaction, do arise [see also

Bijsterbosch and Lyklema (1978)]. Such intermediate situations may also arise as a result of ionic association/dissociation phenomena on the surfaces which can result in surface potentials and/or surface charge densities which do not remain constant during interactions. Additional discussions of charge association/dissociation equilibria and some selected equations for the interaction potentials are presented in Rajagopalan and Kim (1981) and references therein. Some more recent references on double layer interaction potentials are given in Chapter 4, Section A.3.

A.4 Electrokinetic Phenomena

Before leaving this discussion of electric double layers, a few remarks with respect to electrokinetic phenomena are appropriate. Briefly, electrokinetic phenomena are those phenomena associated with attempts to shear off the so-called 'mobile' part of the electric double layer of a surface. For general reviews of these phenomena, see Overbeek (1952), Kruyt (1952), Dukhin and Derjaguin (1974), and texts cited previously in this Section. These electrokinetic phenomena may be classified into four types: (1) electrophoresis; (2) electroosmosis; (3) streaming potential; and (4) sedimentation potential. Relevant experiments and experimental results on colloidal systems are described in the books by Shaw (1969), Smith (1976) and, more recently, in reviews by Dukhin and Derjaguin (1974) and in James (1979). Characterization of aqueous colloids by their electrical double layers and surface chemical properties has been reviewed recently by James and Parks (1982).

B. Attractive Surface Forces

This section will be concerned with the dispersion forces which are exerted by all surfaces. These dispersion forces, or London-van der Waals forces, are attractive forces which exist between atoms, molecules, ions and other surfaces and these give rise to attractive (usually) forces between particles and/or surfaces in proximity. The London-van der Waals forces are a result of the fluctuations in the charge density of the electron cloud surrounding the nucleus. These fluctuations, in turn, produce an instantaneous dipole moment which produces an electric field [London (1930); Israelachvili (1974a)]. London (1930) first explained these dispersion forces and derived an expression for the interaction energy for the case of two identical atoms, at some distance x apart, whose interactions are assumed to occur only through the electrons in s-orbitals (only a single electron is involved in the fluctuations in London's model). As well-known, results indicated that the attractive energy between molecules varies inversely with the sixth power of the intermolecular distance.

It should be noted that there are two limitations to the London analysis above. These limitations occur when: (1) molecules, or atoms, are separated by a large distance; in this case, a retardation effect is important; and (2) separation between atoms or molecules is very small, on the order of the diameter of the atoms or molecules themselves. When atomic contact occurs, for example, the interacting atoms can no longer be treated as simple dipoles, and methods other than London's must be used for calculating the interaction energy. The theory of "retarded" van der Waals forces has been considered by Casimir and Polder (1948); additional details on this topic are presented in Tabor (1979) and in a recent review by Tabor

(1982). For the situation of very small separations (e.g., atomic contact), two approaches to calculate the correct interaction energy have been proposed. The first treats the interacting atoms not as simple dipoles, but rather as quadrupoles and octupoles [see Buckingham (1978)]. The second method for including these short-range forces in the interaction energy is due to Gordon and Kim (1972). Gordon and Kim treat the electron density as the sum of the densities of the original, unperturbed atoms. For additional details, refer to the above sources; in addition, see the review by Clugston (1978).

The forces between two macroscopic bodies which arise from dispersion effects are usually referred to as surface forces. These London-van der Waals forces between macroscopic bodies were explained quantitatively by Hamaker (1937), and subsequently by Lifshitz (1955, 1956). The theory of Hamaker is based on the assumption that the attractive force between two particles, each consisting of a large number of molecules, is simply the sum of the interactions between all pairs of molecules on different particles. The resultant attractive energy, in the case of parallel flat media, varies inversely with the square of the distance between the media. The Hamaker theory, however, is not adequate to completely describe the van der Waals phenomenon. Lifshitz (1955, 1956) and Dzyaloshinskii, Lifshitz and Pitaevskii (1961) developed a more accurate theory to describe the dispersion forces between macroscopic bodies; Lifshitz treats the bodies and the intervening medium in terms of their bulk properties. Details of these two methods are described in many review articles and texts including Gregory (1969), Israelachvili and Tabor (1973), Israelachvili (1974a,b) Langbein (1974), Mahanty and Ninham (1976), Margenau and Kestner (1971), and Parsegian (1975). The values of the Hamaker coefficients of many metals and organic and inorganic compounds needed for the calculations of dispersion forces can be found in the reviews of Visser (1972, 1976). Approximate expressions for the retarded van der Waals interactions were derived by Schenkel and Kitchener (1960); other approximate expressions are also given in Clayfield and Lumb (1966), Ho and Higuchi (1968) and Czarnecki (1979). Exact expressions for the retarded van der Waals interactions are also available [see Clayfield, Lumb and Mackey (1971)].

The van der Waals interactions are reasonably well-understood for separations in the range of about 1 to 10 nm [i.e., unretarded forces], and for separations from 10 up to about 1000 nm [i.e., retarded forces]. The Lifshitz, or continuum, approach and expressions developed on that basis are reasonably accurate in those ranges. Experimental measurements of surface forces (both unretarded and retarded behavior) have been made by several groups. In particular, measurements of forces between mica surfaces in air as a function of separation distance have been performed by Tabor and Israelachvili and co-workers [e.g., Tabor (1969, 1977); Tabor and Winterton (1969); Israelachvili (1978); Israelachvili and Adams (1978); Israelachvili and Tabor (1972, 1973); earlier experiments are referenced in several of these works]. Recently, Israelachvili and co-workers have measured the surface forces between mica surfaces immersed in liquids [Horn and Israelachvili (1980) and Horn, Israelachvili and Perez (1981); see also Israelachvili and Ninham (1977)]. These results, moreover, are significant to investigations into the structure of liquids. Techniques for the measurement of these forces and other forces between solids are discussed in detail in the recent review by Lodge (1983).

At very small separations [i.e., less than about 0.2 nm] or in situations of atomic contact, neither the Lifshitz nor the Hamaker theories are accurate and generally underestimate the magnitude of the interparticle forces. These short-ranged forces at atomic contact may not be of primary significance in typical colloidal systems (where particles are usually separated by distances on the order of tens of nanometers or more); however, they have a fundamental role in such phenomena as the adhesion between particles (in the cases where molecular contact might occur), and in surface energies. The surface forces and adhesion between a _rigid_ sphere and a _rigid_ flat surface is perhaps the simplest case of interest for colloidal systems. Derjaguin [Derjaguin, Muller and Toporov (1975)], however, modified this undeformable solids case to account for elastic deformation of the bodies upon contact. Another analysis, based on assuming an elastic sphere but a rigid flat surface has been developed by Johnson, Kendall and Roberts (1971). Descriptions of these models and interpretations of the results are presented in Tabor (1977, 1982).

Another approach to the analysis of adhesion is based on fracture mechanics; this approach is particularly useful for such substances as polymers or rubber or other viscoelastic solids [see, for example, Greenwood and Johnson (1981)]. Details of this approach are available in Tabor (1977, 1982), and references therein.

Negative van der Waals Forces

As described in the preceeding discussion, van der Waals forces play an important role in many particle-particle or particle-surface interactions. Usually, these van der Waals forces are _attractive_ in nature; however, it is also possible that the van der Waals interactions between two different materials immersed in a liquid medium can be repulsive. Before leaving this section, a few comments on such repulsive van der Waals forces are presented below.

The possibility that the van der Waals interactions between different particles or molecules in a liquid can be attractive, repulsive or zero was suggested in Hamaker's paper on London-van der Waals forces [Hamaker (1937)]. Specifically, if the effective Hamaker coefficient of the system consisting of two different materials suspended in a liquid medium is negative, then the net van der Waals interaction for the system is repulsive. This situation, in turn, occurs when the magnitude of the Hamaker coefficient of the liquid medium is intermediate between the values of the Hamaker coefficients of the two different interacting bodies. [See, for example, Neumann, Omenyi and van Oss (1979); van Oss, Absolom and Neumann (1980, 1982); van Oss, Omenyi and Neumann (1979); and the various references listed in these.]

The implications of negative effective Hamaker coefficients and the subsequent repulsive van der Waals interactions with respect to physical phenomena in colloidal systems such as phase changes, stability and others have been considered by several investigators, in particular, van Oss and co-workers. The realization that the effective Hamaker coefficient can be positive, negative or zero was discussed, from a theoretical point of view, by Visser (1972) in his review of equations for calculating these Hamaker coefficients. More recently, Visser has reviewed and summarized the history and status of the subject of negative Hamaker coefficients [Visser (1981)]. Experimental techniques and experimental evidence of the existence and effects of negative Hamaker coefficients are described in the recent article and review by van Oss et al. (1983); earlier experiments by other scientists which confirmed this phenomenon [e.g., Derjaguin, Zheleznyi, and Tkachev (1972); Sonntag, Buske and Fruhner (1972); Wittman, Splitgerber and Ebert (1971); Churaev (1974); Kruslyakov (1974); among others] are reviewed and discussed in Visser (1981) and van Oss et al. (1983). Applications of the phenomenon of repulsive van der Waals interactions to several separation methods are possible; for example, affinity chromatography [see van Oss et al. (1978, 1979, 1983) and references therein; see also van Oss, Absolom and Neumann (1982)]; hydrophobic chromatography [see van Oss, Absolom and Neumann (1979); van Oss et al. (1978); Barford et al. (1982)]; and other applications [e.g., van Oss et al. (1983); and Absolom, van Oss and Neumann (1981)].

C. Structural Forces

These are strong repulsive forces which are important over very short ranges. These structural forces [also referred to sometimes as solvent mediated forces] arise as a result of changes in the solvent structure (i.e., changes in the local ordering of molecules) in the vicinity of a surface or interface. Such changes arise as a result of packing constraints in interfacial regions, for example. These forces have received increasing attention from researchers during the last five to ten years [see, for example, Ash, Everett and Rader (1973); Chan et al. (1978, 1979, 1980); Israelachvili (1978); Israelachvili and Adams (1978); Israelachvili and Ninham (1977); Lane and Spurling (1979, 1980); Marcelja and Radić (1976); Mitchell, Ninham and Pailthorpe (1977, 1978a,b,c); Snook and van Megen (1979, 1980, 1981); and van Megen and Snook (1979)]. In the case of water as solvent, a phenomenological theory of repulsive structural forces between surfaces due to the ordering of water molecules at an interface was first proposed by LeNeveu et al. (1975), LeNeveu, Rand and Parsegian (1976), and Marcelja and Radić (1976).

Structural forces may exert a relatively large effect for small separation distances between the particles or surfaces; however, it is believed that these do not cause a significant addition to the long-range tail of the forces but may be of great relevance to determining colloidal stability. At small separations, though, the structural forces may be important to such phenomena as repeptization, i.e., resuspension of a coagulated dispersion. [The nature of these structural forces at small separation distances is oscillatory (see Figure 3.C.1), with the wavelength of the oscillation being on the order of the distance of separation between

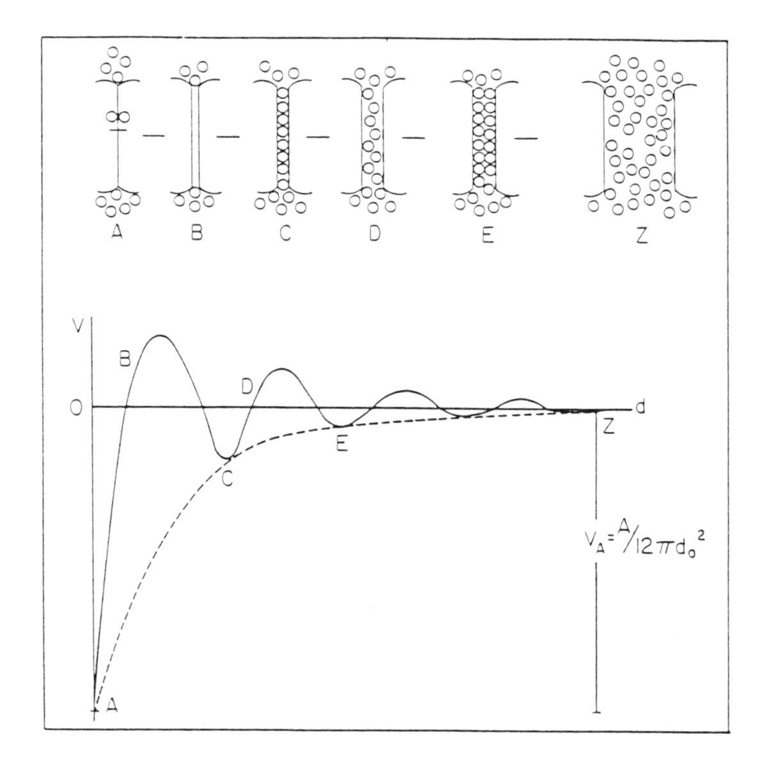

Figure 3.C.1 Schematic representation of structural forces in the vicinity of a surface [from Ninham (1980)].

The parameter V is the interaction energy, A is the net Hamaker constant, and d_0 is the diameter of <u>all</u> the atoms considered in the interaction (solid as well as liquid).

the solvent molecules near the interface.] For extremely small separation distances, the nature of the structural forces may be either repulsive or attractive, as a result of the ease (or difficulty) of removing the last layer of molecules between two surfaces [Overbeek (1982)]. This subject needs additional attention. In general, the area of structural forces is one in which rapid advances are being made and which is attracting increased attention from colloid scientists (see, also, Chapter 4, Section A.5).

D. Forces Relevant to Biological Systems

Roughly speaking, the attachment of a particle to a surface may be viewed as involving two steps: a transport step and an adhesion/adsorption step. Both of these steps are influenced by diverse physical, chemical and colloidal forces. Moreover, in the case where the particle or surface is a living organism, biological forces will be important. The biological nature of the particle or surface will have a role in both steps of the attachment process. Lips and Jessup (1979), for example, discuss the physical forces which influence the attachment of a colloidal particle to a surface in the context of the problem of bacterial adhesion. Other investigators have also considered problems relevant to stability and structure of biological colloids and/or adhesion of microorganisms to surfaces [e.g., see Tanford (1973); Israelachvili and Mitchell (1975); Israelachvili, Mitchell and Ninham (1976); Israelachvili and Ninham (1977); and Jones (1977); see also articles and references in the texts edited by Reissig (1977); Ellwood, Melling and Rutter (1979); Bitton and Marshall (1980); Berkeley et al. (1981); and Cooper and Peppas (1982)].

The primary forces which are important to the transport and adhesion of any particle to a surface can be classified into three categories: (1) forces which arise from the motion of the particle in the fluid, e.g., hydrodynamic and diffusion forces; (2) external forces such as those due to electric, magnetic or gravitational fields; and (3) chemical and colloidal forces which result from interactions of the molecules of the media, e.g., van der Waals forces, coulombic forces, and others. The first two categories are important to the transport step, i.e., in bringing the particle (or organism) to the surface or substrate. Forces in the third class of forces are those which are relevant to the adhesion/adsorption step.

The major colloidal forces considered to be relevant to biological systems are discussed below; these include coulombic forces and van der Waals forces, which were discussed in Sections A and B, respectively, of this Chapter. Additional comments below are addressed specifically to biological systems.

Coulombic Forces

These repulsive electrostatic forces are also important to biological systems since surfaces of microorganisms such as bacteria, cells, etc. generally have a net charge associated with them. For example, surfaces of both eukaryotic and prokaryotic cells have a negative charge associated with them due to ionization of various chemical groups of the surface region [e.g., Jones (1977)]. Attempts to apply the DLVO theory to biological colloids are described, for example, in Lips and Jessup (1979) or Jones (1977), although this theory is inadequate to describe interactions (e.g., cell-cell interactions) at short ranges. As noted in Section A of this Chapter, ionic dissociation/association phenomena on the surfaces may be an important factor in understanding electrostatic interactions in colloidal systems. In particular, for biological systems, surface charges may arise due to dissociation of acidic or basic groups on the surfaces. Determination of electrostatic forces or potential in such cases must account for this [Ninham and Parsegian (1971)]. Other features of the biological surfaces also may affect significantly the electrostatic forces and interactions. For example, the surface charge of biological cells is not associated with an infinitesimally thin layer, but rather with a layer on the order of several nanometers thick [Parsegian and Gingell (1973)].

Other factors which may be important include the effect of the cell interior on electrostatic potential, effects of non-uniformities of charge distribution on the biological surface, e.g., areas of positive charge may exist on the surface [Weiss (1974)], effects due to specific adsorption phenomena [Okada et al. (1974)], and effects due to the nature of chemical bonds (in particular, hydrophobic bonds) on interactions [Marshall and Cruickshank (1973); van Oss and Gillman (1972a,b); van Oss, Gillman and Neumann (1975); for a discussion and additional references, see the article by Jones (1977)].

London-van der Waals Forces

The two approaches for calculating the van der Waals forces, the method of Hamaker and the method of Lifshitz, were discussed above. Hamaker (1937) considers the material property, now known as the Hamaker 'constant', to be a constant, dependent only on the material involved. The Lifshitz theory, alternatively, considers this property to be also dependent on the distance of separation and temperature. In general, the Hamaker _function_, and hence the dispersion energy, will reflect contributions from phenomena over the entire frequency spectrum. In inert, non-biological colloidal systems, contributions to the dispersion energy are due largely to fluctuations in the visible and ultraviolet regions. In biological systems, however, contributions from low frequency components may be significant [Parsegian and Ninham (1970); Parsegian and Gingell (1973); Parsegian (1973)]. For example, Parsegian and Ninham (1970) found that for hydrocarbon layers or cell walls interacting in water, the "zero frequency" contribution to the dispersion energy was temperature dependent and on the order of sixty percent of the total. The low frequency component is sensitive to electrolyte concentrations, and the net effects of these and other variables need additional investigation.

Finally, as noted in Section B of this Chapter, the effective Hamaker coefficient for an interacting system which consists of two materials in a liquid can be negative, under certain conditions, thereby leading to repulsive van der Waals interactions. This can occur for interactions in biological colloidal systems, e.g., for two different types of cells in a liquid. The influence of repulsive (and/or attractive) van der Waals forces on cell (and other biological materials) interactions has been studied extensively by van Oss and Neumann and co-workers [see, for example, Neumann et al. (1974); van Oss, Gillman and Neumann (1975); van Oss (1978); van Oss et al. (1980); additional references are given in the last].

Other Forces

Other forces which may affect significantly the total interaction energy in biocolloidal systems include structural forces (see Section C of this Chapter) and forces due to adsorbed layers. The latter are particularly relevant to biological and nonaqueous systems in which macromolecular species are present. These 'adsorbed-layer-mediated' forces include: (1) increases in attractive van der Waals forces between two approaching surfaces due to polymer bridging; and (2) steric forces (see Section E of this Chapter). The steric forces can be large relative to the electrostatic forces involved [see, for example, Greig and Jones (1976)]. In general, these effects can occur simultaneously; e.g., steric forces (and accompanying repulsion) and bridging (with its accompanying attraction). There are theories to assess each of these latter two effects in isolation, but not to deal with both effects occurring simultaneously [see Sato and

Ruch (1980) and references therein].

Hydrodynamic Forces and Diffusion Forces

These forces, as mentioned above, are important in the transport step of the adsorption phenomenon. A comprehensive review and discussion of these forces is presented in Rajagopalan and Tien (1979). In the context of biological systems, hydrodynamic forces are particularly important since these may easily exceed the surface forces which cause the microorganism to remain attached (against thermal forces) near the secondary minimum [Lips and Jessup (1979); see also Dahneke (1975)]. This possible effect of interfacial convection is discussed further in Chapter 5.

Deformation Due to Contact

The magnitude and extent of the deformation which occurs when two objects come into contact and adhere will depend upon several factors including the nature of the forces between the two bodies, and the geometries and properties (e.g., elastic behavior) of the objects. [See Section B of this Chapter.] In the case of biological systems (e.g., cells), significant contact deformation can occur [Rees, Lloyd and Thom (1977)] and needs to be taken into account in evaluating the net interaction force and energy.

E. Steric Forces

In addition to the repulsion which arises due to electrostatic origin, a second type of repulsion, termed steric repulsion, is important to understanding colloidal interactions; see Figure 3.E.1. Steric repulsion is a long-ranged repulsion which results from polymers adsorbed to the interacting surfaces. Specifically, steric repulsion occurs when long-chain molecules adsorbed to the particle surface repel one another and subsequently cause a large repulsive force between the particles. The physical basis of the steric repulsion between particles arises from two effects: (1) a volume restriction effect arising from the decrease in possible configurations in the region between the two surfaces, and (2) an osmotic effect due to the relatively high concentration of adsorbed molecules or chains in the region between the particles as they approach; see Figure 3.E.2 [see, for example, Napper (1968, 1970); Evans and Napper (1973a,b); Hesselink (1969, 1971); and Hesselink, Vrij and Overbeek (1971)]. Reviews of steric forces and their role in polymeric stabilization (and extensive bibliographies) are given in Napper (1977) and, more recently, in Napper (1982). The text by Sato and Ruch (1980) presents a detailed discussion of steric stabilization, and other aspects of polymer adsorption. For a discussion of steric stabilization and colloidal systems, see Chapter 4, on Colloidal Stability.

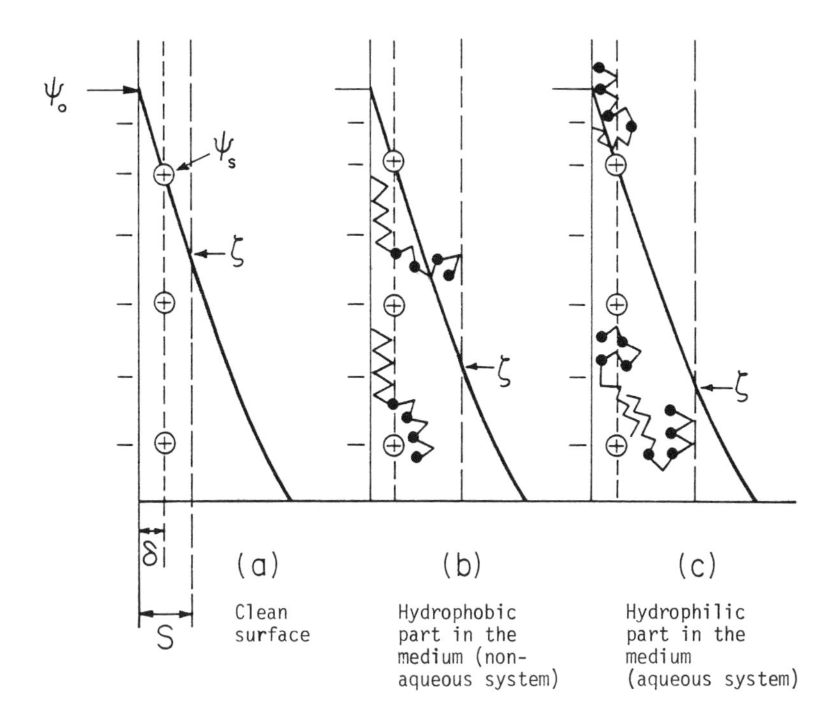

Figure 3.E.1 Effect of nonionic surfactants on a charged surface [from Sato and Ruch (1980)].

The parameter ψ_o = surface potential; ψ_S = Stern potential (at the Stern plane); ζ = zeta potential (at the shear plane). The effect of adsorbed layer becomes important when its thickness is about the same as that of the double layer.

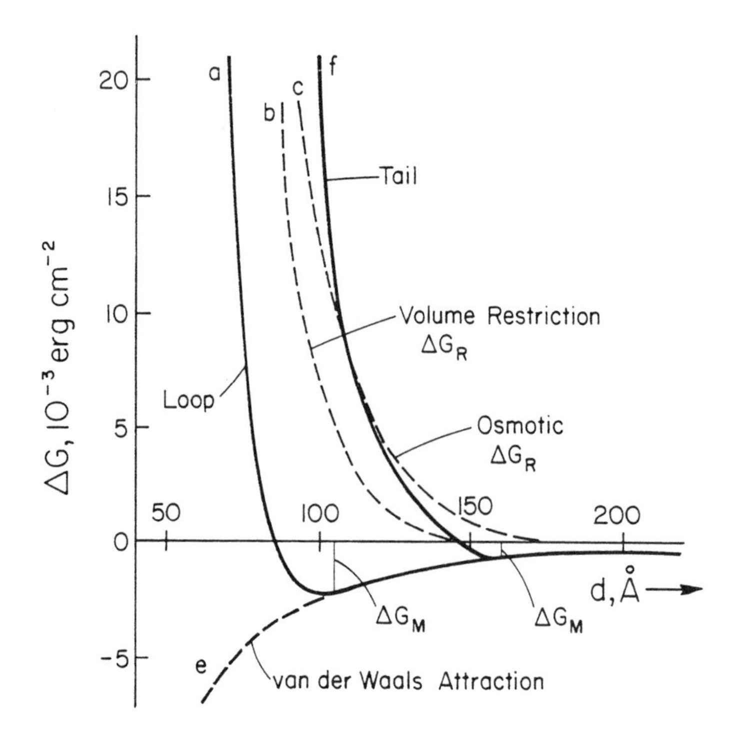

Figure 3.E.2 Total free energy of interaction for particles covered by equal number of adsorbed tails (f) and equal number of loops (a); see Sato and Ruch (1980) for details.

These values are obtained as the sum of van der Waals, volume restriction and osmotic effects. For example, b and c are for equal tails, and b+c+e = f.

Research Needs

In general, there are uncertainties in issues related to the nature and characterization of the various intermolecular and electrokinetic forces which complicate the applications of these forces to colloidal phenomena such as stability and/or other interactions. In particular, the following problems remain unsolved at present, and require additional attention and study.

- The structure of the electric double layer is one outstanding problem in this area. This problem can be further sub-divided into two categories of research needs; one, a need for better understanding of the non-diffuse portion of the electric double layer (i.e., the Stern Layer). Secondly, understanding of the structure of the diffuse layer presents a challenge, especially in the case of asymmetric electrolytes. In the first category, for example, under certain conditions (e.g., near the limit of destabilization), a significant fraction of the total potential decrease is now known to occur in the non-diffuse layer. In the latter category, the validity of the Poisson-Boltzmann equation, or its modifications, requires further study. Computer experiments seem to offer a promising technique for these latter investigations.

- The exact nature, or mode, of the interaction between interacting double layers (e.g., constant potential or constant charge or a mixture of these) is another challenging problem of colloid science. The question as to which mechanism is occurring during dynamic double layer interactions is as yet unresolved. Thus, additional studies on the dynamics of double layer interactions, including the examination of the rates of double layer equilibration versus rates of approach to the surfaces, etc., are necessary.

- Studies of the double layers associated with colloidal species and of electrokinetic phenomena have been concerned primarily with aqueous colloids. There is, thus, a great need to examine the nature of the double layer, its structure and dynamics, and to investigate the nature of electrokinetic phenomena of colloidal particles in non-aqueous media. Very few studies to date have dealt with non-polar media. This is especially true for concentrated dispersions in non-aqueous media since, in such cases, the average interparticle separation is of the same order of magnitude as the double layer thickness (see Chapter 6).

- Techniques for the measurement of zeta potential and electrokinetic properties of very small suspended particles are readily available and are well-developed. However, methods to determine zeta potentials or other electrokinetic properties of large surfaces (e.g., flat plates) are needed.

- Additional experimental studies on the direct measurement of van der Waals forces would reduce the uncertainties which are associated with the estimation of Hamaker constants and with respect to expressions for attractive forces. Related research needs include the development of techniques to measure interaction forces in the presence of electrolytes, or in the presence of fluids.

- In recent years, there has been an increasing amount of attention and interest in the short-range structural forces. Much remains to be learned about the nature and characterization of these structural forces, especially at very small distances of separation of the surfaces. For example, the measurement and role of these short-range forces in repulsive interactions between surfaces is a challenging problem. In particular, the short-range interactions which occur when particles are at atomic contact cannot be adequately explained by the Lifshitz theory. Studies to examine these are required.

- In general, any discussion of the forces relevant to biocolloidal systems is inseparable from that of the forces relevant to nonbiological colloids. The application of colloid science to biological systems, however, is only beginning to be explored. The recent surge of interest in applications of principles of colloid science has focused interest on the colloidal aspects of these systems while, at the same time, drawing attention to the complications that biological factors may introduce. For example, research problems include such issues as the role of polymers in cell adhesion, changes in adsorption and/or membrane composition which accompany biological adsorption, and others (see Section 3.D).

- A key problem in the understanding and application of steric forces in colloidal interactions is that of polymer conformation at interfaces. Direct measurements of steric forces using mica sheets coated with polymer have resulted in some progress in understanding the interactions between steric barriers. Additional studies of this type are needed (see also the comments on Research Needs in Chapter 4).

REFERENCES: SECTION A. ELECTRIC DOUBLE LAYER STRUCTURE AND FORCES

Bell, G. M. and Peterson, G. C., J. Colloid Interface Sci. 41, 542 (1972).

Bijsterbosch, B. H. and Lyklema, J., Adv. Colloid Interface Sci. 9, 147 (1978).

Bockris, J' O. M., Conway, B. E. and Yeager, E. (eds.), The Electric Double Layer, Vol. 1, Comprehensive Treatise of Electrochemistry, Plenum, New York, NY, 1980.

Carnie, S. L., Chan, D. Y. C., Mitchell, D. J. and Ninham, B. W., J. Chem. Phys. 74, 1472 (1981).

Chapman, D. L., Phil. Mag. 25, 475 (1913).

Croxton, T. L. and McQuarrie, D. A., Chem. Phys. Lett. 68, 489 (1979).

Croxton, T. L. and McQuarrie, D. A., Mol. Phys. 42, 141 (1981).

Croxton, T. L., McQuarrie, D. A., Patey, G. N., Torrie, G. M. and Valleau, J.P., Can. J. Chem. 59, 1998 (1981).

Dukhin, S. S. and Derjaguin, B. V., pp. 1-335 in Matijević (1974).

Frens, G., The Reversibility of Irreversible Colloids, Thesis, Utrecht, The Netherlands, 1968.

Good, R. J. and Stromberg, R. R. (eds.), Surface and Colloid Science, Vol. 11, Experimental Methods, Plenum, New York, NY, 1979.

Goodwin, J. W. (ed.), Colloidal Dispersions, The Royal Soc. Chem., London, U. K., 1982.

Gouy, G., J. Phys. 9, 457 (1910).

Grahame, D. C., Chem. Rev. 41, 441 (1947).

Gregory, J., J. Colloid Interface Sci. 51, 44 (1975).

Henderson, D. and Blum, L., J. Chem. Phys. 69, 5441 (1978).

Henderson, D. and Blum, L., Can. J. Chem. 59, 1906 (1981).

Hiemenz, P. C., Principles of Colloid and Surface Chemistry, Marcel Dekker, New York, NY, 1977.

Hogg, R., Healy, T. W. and Fuerstenau, D. W., Trans. Faraday Soc. 62, 1638 (1966).

Honig, E. P. and Mul, P. M., J. Colloid Interface Sci. 36, 258 (1971).

Hunter, R. J., Zeta Potential in Colloid Science, Academic Press, London, U. K., 1981.

James, A. M., pp. 121-185 in Good and Stromberg (1979).

James, R. O. and Parks, G. A., pp. 119-216 in Matijević (1982).

Kar, G., Chander, S. and Mika, T. S., J. Colloid Interface Sci. 44, 347 (1973).

Kerker, M. (ed.), Surface Chemistry and Colloids, Butterworths, London, U. K., 1972.

Kruyt, H. R. (ed.), Colloid Science, Volume 1: Irreversible Systems, Elsevier, Amsterdam, The Netherlands, 1952.

Krylov, V. S. and Levich, V. G., Russian J. Phys. Chem. 37, 1224 (1963).

Levine, S. and Bell, G. M., Discuss. Faraday Soc. 42, 69 (1966).

Levine, S. and Outhwaite, C. W., J. Chem. Soc. Faraday Trans. II 74, 1670 (1978).

Levine, S., Outhwaite, C. W. and Bhuiyan, L. B., J. Electroanal. Chem. 123, 105 (1981).

Loeb, A. L., Overbeek, J. Th. G. and Wiersema, P. H., The Electrical Double Layer Around A Spherical Colloid Particle, M. I. T. Press, Cambridge, Mass., 1961.

Lyklema, J., pp. 47-70 in Goodwin (1982).

Matijević, E. (ed.), Surface and Colloid Science, Vol. 7, Wiley-Interscience, New York, NY, 1974.

Matijević, E. (ed.), Surface and Colloid Science, Vol. 12, Plenum, New York, NY, 1982.

Napper, D. H. and Hunter, R. J., pp. 241-306 in Kerker (1972).

Outhwaite, C. W. and Bhuiyan, L. B., J. Chem. Soc. Faraday Trans. II 78, 775 (1982).

Outhwaite, C. W. and Bhuiyan, L. B., J. Chem. Soc. Faraday Trans. II 79, 707 (1983).

Overbeek, J. Th. G., pp. 302-341 in Kruyt (1952).

Rajagopalan, R. and Kim, J. S., J. Colloid Interface Sci. 83, 428 (1981).

Shaw, D. J., Electrophoresis, Academic Press, New York, NY, 1969.

Shaw, D. J., Introduction to Colloid and Surface Chemistry, 3rd ed., Butterworths, London, U. K., 1980.

Smith, I., Zone Electrophoresis, Heinemann, London, U. K., 1976.

Snook, I. and van Megen, W., J. Chem. Phys. 75, 4104 (1981).

Stern, O., Z. Elektrochem. 30, 508 (1924).

Torrie, G. M. and Valleau, J. P., J. Chem. Phys. 86, 3251 (1982).

Torrie, G. M., Valleau, J. P. and Patey, G. N., J. Chem. Phys. 76, 4615 (1982).

van Megen, W. and Snook, I., J. Chem. Phys. 73, 4656 (1980).

Verwey, E. J. W. and Overbeek, J. Th. G., Theory of the Stability of Lyophobic Colloids, Elsevier, Amsterdam, The Netherlands, 1948.

Vold, R. D. and Vold, M. J., Colloid and Interface Chemistry, Addison-Wesley, Reading, Mass., 1983.

Wiese, G. R. and Healy, T. W., Trans. Faraday Soc. 66, 490 (1970).

REFERENCES: SECTION B. ATTRACTIVE SURFACE FORCES

Absolom, D. R., van Oss, C. J. and Neumann, A. W., Transfusion 21, 663 (1981).

Barford, R. A., Sliwinski, B. J., Breyer, A. C. and Rothbart, H. L., J. Chromatog. 235, 281 (1982).

Buckingham, A. D., pp. 1-67 in Pullman (1978).

Casimir, H. B. G. and Polder, D., Phys. Rev. 73, 360 (1948).

Churaev, N. V., Coll. J. USSR 36, 287 (1974).

Clayfield, E. J. and Lumb, E. C., Discuss. Faraday Soc. 42, 285 (1966).

Clayfield, E. J., Lumb, E. C. and Mackey, P. H., J. Colloid Interface Sci. 37, 382 (1971).

Clugston, M. J., Adv. Phys. 27, 893 (1978).

Czarnecki, J., J. Colloid Interface Sci. 72, 361 (1979).

Danielli, J. F., Rosenberg, M. D. and Cadenhead, D. A. (eds.), Progress in Surface and Membrane Science, Vol. 7, Academic Press, New York, NY, 1973.

Derjaguin, B. V., Muller, V. M. and Toporov, Y. P., J. Colloid Interface Sci. 53, 314 (1975).

Derjaguin, B. V., Zheleznyi, B. V. and Tkachev, A. P., Dokl. Akad. Nauk. SSSR 206, 1146 (1972).

Dzayaloshinskii, I. E., Lifshitz, E. M. and Pitaevskii, L. P., Adv. Phys. 10, 165 (1961).

Goodwin, J. W. (ed.), Colloidal Dispersions, The Royal Soc. Chem., London U. K., 1982.

Gordon, R. G. and Kim, Y. S., J. Chem. Phys. 56, 3122 (1972).

Greenwood, J. A. and Johnson, K. L., Phil. Mag. 43, 697 (1981).

Gregory, J., Adv. Colloid Interface Sci. 2, 396 (1969).

Gribnau, T. C. J., Visser, J. and Nivard, R. J. F. (eds.), Affinity Chromatography and Related Techniques, Elsevier, Amsterdam, The Netherlands, 1982.

Hamaker, H. C., Physica 4, 1058 (1937).

Ho, N. F. H. and Higuchi, W. I., J. Pharm. Sci. 57, 436 (1968).

Horn, R. G. and Israelachvili, J. N., Chem. Phys. Lett. 71, 192 (1980).

Horn, R. G., Israelachvili, J. N. and Perez, E., J. Physique 42, 23 (1981).

Israelachvili, J. N., Contemp. Phys. 15, 159 (1974a).

Israelachvili, J. N., Quart. Rev. Biophys. 6, 341 (1974b).

Israelachvili, J. N., Discuss. Faraday Soc. 65, 20 (1978).

Israelachvili, J. N. and Adams, G. E., J. Chem. Soc. Faraday Trans. I 74, 975 (1978).

Israelachvili, J. N. and Ninham, B. W., pp. 15-26 in Kerker, Zettlemoyer and Rowell (1977).

Israelachvili, J. N. and Tabor, D., Proc. Roy. Soc. A 331, 19 (1972).

Israelachvili, J. N. and Tabor, D., pp. 1-55 in Danielli, Rosenberg, and Cadenhead (1973).

Johnson, K. L., Kendall, K. and Roberts A. D., Proc. Roy. Soc. A 324, 301 (1971).

Kerker, M., Zettlemoyer, A. C. and Rowell, R. L. (eds.), Colloid and Interface Science, Vol. 1, Academic Press, New York, NY, 1977.

Kruslyakov, P. M., Coll. J. USSR 36, 145 (1974).

Langbein, D., Theory of van der Waals Attraction, Springer-Verlag, Berlin, Germany, 1974.

Lifshitz, E. M., J. Exp. Theor. Phys. USSR 29, 94 (1955).

Lifshitz, E. M., Sov. Phys. JETP 2, 73 (1956).

Lodge, K. B., Adv. Colloid Interface Sci. 19, 27 (1983).

London, F., Zeit. Phys. 63, 245 (1930).

Mahanty, J. and Ninham, B. W., Dispersion Forces, Academic Press, London, U. K., 1976.

Margenau, H. and Kestner, N. R., Theory of Intermolecular Forces, Pergamon, Oxford, U. K., 1971.

Matijević, E. (ed.), Surface and Colloid Science, Vol. 8, Wiley-Interscience, New York, NY, 1976.

Neumann, A. W., Omenyi, S. N. and van Oss, C. J., Colloid Polymer Sci. 257, 413 (1979).

Parsegian, V. A., pp. 27-72 in van Olphen and Mysels (1975).

Pullman, B. (ed.), Intermolecular Interactions: From Diatomics to Polymers, Wiley, Chichester, U. K., 1978.

Schenkel, J. H. and Kitchener, J. A., Trans. Faraday Soc. 56, 161 (1960).

Sonntag, H., Buske, N. and Fruhner, M., Kolloid Z. Z. Pol. 250, 330 (1972).

Tabor, D., J. Colloid Interface Sci. 31, 364 (1969).

Tabor, D., pp. 3-14 in Kerker, Zettlemoyer and Rowell (1977).

Tabor, D., Solids, Liquids and Gases, Cambridge Univ. Press, Cambridge, U. K., 1979.

Tabor, D., pp. 23-46 in Goodwin (1982).

Tabor, D. and Winterton, R. H. S., Proc. Roy. Soc. A 312, 435 (1969).

van Olphen, H. and Mysels, K. J. (eds.), Physical Chemistry: Enriching Topics from Colloid and Surface Science, IUPAC Commission I.6, Theorex, La Jolla, CA, 1975.

van Oss, C. J., Absolom, D. R., Grossberg, A. L. and Neumann, A. W., Immunol. Commun. 8, 11 (1979).

van Oss, C. J., Absolom, D. R. and Neumann, A. W., Separ. Sci. Technol. 14, 305 (1979).

van Oss, C. J., Absolom, D. R. and Neumann, A. W., Colloids Surfaces 1, 45 (1980).

van Oss, C. J., Absolom, D. R. and Neumann, A. W., pp. 29-37 in Gribnau, Visser and Nivard (1982).

van Oss, C. J., Neumann, A. W., Omenyi, S. N. and Absolom, D. R., Separ. Purif. Meth. 7, 245 (1978).

van Oss, C. J., Omenyi, S. N. and Neumann, A. W., Colloid Polymer Sci. 257, 737 (1979).

van Oss, C. J., Visser, J., Absolom, D. R., Omenyi, S. N. and Neumann, A. W., Adv. Colloid Interface Sci. 18, 133 (1983).

Visser, J., Adv. Colloid Interface Sci. 3, 331 (1972).

Visser, J., pp. 3-84 in Matijević (1976).

Visser, J., Adv. Colloid Interface Sci. 15, 157 (1981).

Wittmann, F., Splitgerber, H. and Ebert, K., Z. Phys. 245, 354 (1971).

REFERENCES: SECTION C. STRUCTURAL FORCES

Ash, S. G., Everett, D. H. and Rader, C. J., J. Chem. Soc. Faraday Trans. II 69, 1256 (1973).

Chan, D. Y. C., Mitchell, D. J., Ninham, B. W. and Pailthorpe, B. A., Mol. Phys. 35, 1669 (1978).

Chan, D. Y. C., Mitchell, D. J., Ninham, B. W., J. Chem. Soc. Faraday Trans. II 75, 556 (1979).

Chan, D. Y. C., Mitchell, D. J., Ninham, B. W. and Pailthrope, B. A., J. Chem. Soc. Faraday Trans. II 76, 776 (1980).

Goodwin, J. W. (ed.), Colloidal Dispersions, The Royal Soc. Chem., London, U. K., 1982.

Israelachvili, J. N., Discuss. Faraday Soc. 65, 20 (1978).

Israelachvili, J. N. and Adams, G. E., J. Chem. Soc. Faraday Trans. I 74, 975 (1978).

Israelachvili, J. N. and Ninham, B. W., pp. 15-26 in Kerker, Zettlemoyer and Rowell (1977).

Kerker, M., Zettlemoyer, A. C. and Rowell, R. L. (eds.), Colloid and Interface Science, Vol. 1, Academic Press, New York, NY, 1977.

Lane, J. E. and Spurling, T. H., Chem. Phys. Lett. 67, 107 (1979).

Lane, J. E. and Spurling, T. H., Aust. J. Chem. 33, 231 (1980).

LeNeveu, D. M., Rand, R. P., Gingell, D. and Parsegian, V. A., Science 191, 399 (1975).

LeNeveu, D. M., Rand, R. P. and Parsegian, V. A., Nature 259, 601 (1976).

Marcelja, S. and Radić, N., Chem. Phys. Lett. 42, 129 (1976).

Mitchell, D. J., Ninham, B. W. and Pailthorpe, B. A., Chem. Phys. Lett. 51, 257 (1977).

Mitchell, D. J., Ninham, B. W. and Pailthorpe, B. A., J. Colloid Interface Sci. 64, 194 (1978a).

Mitchell, D. J., Ninham, B. W. and Pailthorpe, B. A., J. Chem. Soc. Faraday Trans. II 74, 1098 (1978b).

Mitchell, D. J., Ninham, B. W. and Pailthorpe, B. A., J. Chem. Soc. Faraday Trans. II 74, 1116 (1978c).

Ninham, B. W., J. Phys. Chem. 84, 1423 (1980).

Overbeek, J. Th. G., pp. 1-22 in Goodwin (1982),

Snook, I. K. and van Megen, W. J., J. Chem. Phys. 70, 3099 (1979).

Snook, I. K. and van Megen, W. J., J. Chem. Phys. 72, 2907 (1980).

Snook, I. K. and van Megen, W. J., J. Chem. Soc. Faraday Trans. II 77, 181 (1981).

van Megen, W. J. and Snook, I. K., J. Chem. Soc. Faraday Trans. II 75, 1095 (1979).

REFERENCES: SECTION D. FORCES RELEVANT TO BIOLOGICAL SYSTEMS

Berkeley, R. C. W., Lynch, J. M., Melling, J., Rutter, P. and Vincent, B. (eds.), Microbial Adhesion to Surfaces, Wiley, New York, NY, 1981.

Bitton, G. and Marshall, K. C. (eds.), Adsorption of Microorganisms to Surfaces, Wiley-Interscience, New York, NY, 1980.

Blank, M. (ed.), Bioelectrochemistry: Ions, Surfaces, Membranes, American Chemical Society, Washington, D. C., 1980.

Cooper, S. L. and Peppas, N. A. (eds.), Biomaterials: Interfacial Phenomena and Applications, American Chemical Society, Washington, D. C., 1982.

Dahneke, B., J. Colloid Interface Sci. 50, 89 (1975).

Ellwood, D. C., Melling, J. and Rutter, P. (eds.), Adhesion of Microorganisms to Surfaces, Academic Press, London, U. K., 1979.

Greig, R. G. and Jones, M. N., J. Theor. Biol. 63, 405 (1976).

Hamaker, H. C., Physica 4, 1058 (1937).

Israelachvili, J. N. and Mitchell, D. J., Biochim. Biophys. Acta 389, 13 (1975).

Israelachvili, J. N., Mitchell, D. J. and Ninham, B. W., J. Chem. Soc. Faraday Trans II 72, 1525 (1976).

Israelachvili, J. N. and Ninham, B. W., pp. 15-26 in Kerker, Zettlemoyer and Rowell (1977).

Jones, G. W., pp. 139-176 in Reissig (1977).

Kerker, M., Zettlemoyer, A. C. and Rowell, R. L. (eds.), Colloid and Interface Science, Vol. 1, Academic Press, New York, NY, 1977.

Lee, L. H. (ed.), Recent Advances in Adhesion, Gordon and Breach, London, U. K., 1973.

Lips, A. and Jessup, N. E., pp. 5-27 in Ellwood, Melling and Rutter (1979).

Marshall, K. C. and Cruickshank, R. H., Arch. Microbiol. 91, 29 (1973).

Neumann, A. W., Good, R. J., Hope, C. J. and Sejpal, M., J. Colloid Interface Sci. 49, 291 (1974).

Ninham, B. W. and Parsegian, V. A., J. Theor. Biol. 31, 405 (1971).

Okada, T. S., Takeichi, T., Yasuda, K. and Ueda, M. J., Adv. Biophys. 6, 157 (1974).

Parsegian, V. A., Ann. Rev. Biophys. Bioeng. 2, 221 (1973).

Parsegian, V. A. and Gingell, D., pp. 153-190 in Lee (1973).

Parsegian, V. A. and Ninham, B. W., Biophysical J. 10, 664 (1970).

Rajagopalan, R. and Tien, C., pp. 179-269 in Wakeman (1979).

Rees, D. A., Lloyd, C. W. and Thom, D., Nature 267, 124 (1977).

Reissig, J. L. (ed.), Microbial Interactions, Chapman and Hall, London, U. K., 1977.

Sato, T. and Ruch, R., Stabilization of Colloidal Dispersions by Polymer Adsorption, Marcel Dekker, New York, NY, 1980.

Tanford, C., The Hydrophobic Effect: Formation of Micelles and Biological Membranes, Wiley, New York, NY, 1973.

van Oss, C. J., Ann. Rev. Microbiol. 32, 19 (1978).

van Oss, C. J. and Gillman, C. F., J. Reticuloendothelial Soc. 12, 283 (1972a).

van Oss, C. J. and Gillman, C. F., J. Reticuloendothelial Soc. 12, 497 (1972b).

van Oss, C. J., Gillman, C. F., and Neumann, A. W., Phagocytic Engulfment and Cell Adhesiveness, Marcel Dekker, New York, NY, 1975.

van Oss, C. J., Neumann, A. W., Good, R. J. and Absolom, D. R., pp. 107-114 in Blank (1980).

Wakeman, R. J. (ed.), Progress in Filtration and Separation, Vol. 1, Elsevier, Amsterdam, The Netherlands, 1979.

Weiss, L., Expt. Cell Res. 83, 311 (1974).

REFERENCES: SECTION E. STERIC FORCES

Evans, R. and Napper, D. H., Kolloid Z. Z. Polym. 251, 329 (1973a).

Evans, R. and Napper, D. H., Kolloid Z. Z. Polym. 251, 409 (1973b).

Goodwin, J. W. (ed.), Colloidal Dispersions, The Royal Soc. Chem., London, U. K., 1982.

Hesselink, F. Th., J. Phys. Chem. 73, 3488 (1969).

Hesselink, F. Th., J. Phys. Chem. 75, 65 (1971).

Hesselink, F. Th., Vrij, A. and Overbeek, J. Th. G., J. Phys. Chem. 75, 2094 (1971).

Napper, D. H., Trans. Faraday Soc. 64, 1701 (1968).

Napper, D. H., J. Colloid Interface Sci. 32, 106 (1970).

Napper, D. H., J. Colloid Interface Sci. 58, 390 (1977).

Napper, D. H., pp. 99-128 in Goodwin (1982).

Sato, T. and Ruch, R., Stabilization of Colloidal Dispersions by Polymer Adsorption, Marcel Dekker, New York, NY, 1980.

PART II

DILUTE DISPERSIONS

4

Colloidal Stability

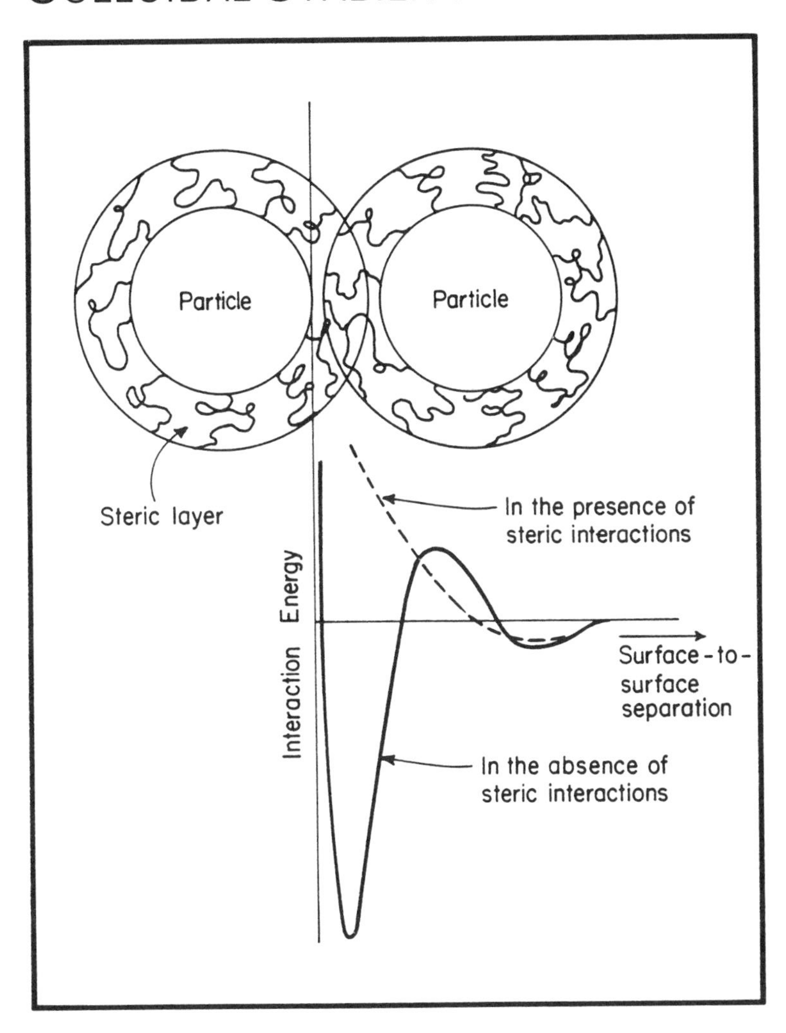

CHAPTER 4

COLLOIDAL STABILITY[1]

Overview

Historically, the need to understand the stability or the instability of colloidal dispersions has formed the central motivating factor in the study of colloidal interactions and in the development of colloid science. The concept of stability in this context is generally understood to mean kinetic stability, i.e., stability imposed by a strong repulsive barrier acting against contact between the suspended particles. Consequently, the colloids of interest here are lyophobic; e.g., latex dispersions, dispersed paints, dispersions used in photographic film, emulsions, gold and silver iodide sols, and the like. [The lyophilic colloids, such as micelles, are thermodynamically (indefinitely) stable; see Chapter 2.]

The London-van der Waals dispersion forces are at the origin of the tendency of lyophobic colloids to coagulate and aggregate. This and the fact that these forces are also crucial for understanding adhesion phenomena have contributed to extensive research on the nature and the origin of these forces and the material constants that are needed to estimate their magnitudes. These are summarized in Chapter 3, Section B.

The source of repulsive forces needed to stabilize the dispersion against the attractive forces is usually one of two kinds:

(i) Coulombic repulsion due to electric charges on the particles or electrostatic interactions between the ionic double layers surrounding the particles, or

---

[1]A slightly revised version of this Chapter is scheduled to appear in Chemical Engineering Communications.

(ii) Steric repulsion introduced by large molecules or polymeric chains adsorbed on the particles.

In view of this, the mechanism of stabilization is classified under the labels 'electrostatic or electrocratic stabilization' or 'steric stabilization', depending on how the kinetic stability is introduced. A multitude of problem areas has been identified in each of these over the years, and current research activities address almost all these individual components; these include the nature of electrical double layer interactions, ionic adsorption/desorption equilibria at interfaces, adsorption and conformation of polymeric molecules at interfaces, etc. It is convenient to review these using the above, logical classification scheme, although some of the problems are necessarily common to both.

A. Electrostatic Stabilization

The Derjaguin-Landau-Verwey-Overbeek theory of electrostatic stabilization (abbreviated commonly as DLVO theory), which attributes the stability (or the lack of it) to the relative strength of electrostatic repulsion over van der Waals attraction, still forms the cornerstone of this area. In addition to the classical and pioneering publications of Derjaguin and Landau [Derjaguin (1940a,b); Derjaguin and Landau (1941a,b, 1945)] and Verwey and Overbeek [Verwey (1942, 1945); Verwey and Overbeek (1946, 1948)], numerous lucid reviews of recent vintage are available on the achievements, strengths and weaknesses of this theory [Overbeek (1977, 1982); Ottewill (1977)]. These latter reviews and most of the standard textbooks on colloids [see Shaw (1980); Hiemenz (1977); Vold and Vold (1983)] outline the basic electrochemical concepts, transport equations [e.g., Smoluchowski equation; Smoluchowski (1916a,b, 1917)], and standard experimental techniques used in the study of colloid stability; therefore, we shall confine the following discussion to a review of the major directions the research in this area has taken and the types of questions that have been investigated.

The major directions in research have been the following.

A.1 Experimental Verification of Rate of Coagulation

The initial attempts in the investigations of colloid stability were primarily concerned with measuring the stability ratios for various dispersions at a variety of electrochemical conditions. The stability ratio is defined as the ratio of collision rate between the particles when there is no potential barrier against collision [that is, at *rapid* coagulation; Smoluchowski (1916a,b, 1917)] to the collision rate in the presence of the barrier [*slow* coagulation; Fuchs (1934)]. In addition, the transition from slow coagulation to rapid coagulation as the electrolyte concentration increases is very sudden, and the electrolyte concentration at which this happens is called the critical coagulation concentration. The initial experiments focused on these two aspects of the problem. The literature on these is very extensive, but the general conclusions are the same:

(i) the rapid coagulation rate, under favorable electrochemical conditions, is given by Smoluchowski's results within experimental accuracy,

(ii) the slopes of log-log plots of stability ratios versus electrolyte concentrations have the correct order of magnitude,

(iii) the critical coagulation concentration is sharp because of the sensitivity of the potential barrier to the electrolyte concentration,

(iv) the critical coagulation concentration decreases with increases in the valence of counterions in qualitative agreement with the Schulze-Hardy rule [Schulze (1882, 1883); Hardy (1900a,b)], and

(v) the experimental stability ratios do *not* show the dependence on the particle radius predicted by theory [Ottewill and Shaw (1966)].

The lack of complete agreement with the theory noted in the last two conclusions is particularly important since it has led to the recent directions in research on colloid stability (see Sections A.2 - A.4 below).

A.2 Hydrodynamic Retardation Corrections to Collision Rate

In an attempt to explain the discrepancy concerning the lack of sensitivity of stability ratio to particle size, several authors have recalculated the equations for collision rate by properly accounting for the increased hydrodynamic resistance to collisions between the particles [Derjaguin and Muller (1967); Spielman (1970); Honig, Roebersen and Wiersema (1971); Deutch and Felderhof (1973)]. These refinements, motivated by a remark by Derjaguin (1966), have led to an appreciation of the influence of hydrodynamic resistance in coagulation and have been confirmed, within experimental uncertainty, by Lichtenbelt, Ras and Wiersema (1974) and Lichtenbelt, Pathmamanoharan and Wiersema (1974). Nevertheless, the original discrepancy (concerning the dependence of stability on particle size) remains unanswered (see, also, Section C.1 below).

A.3 Dynamics of Interfacial Electrochemistry

Traditionally, studies on colloidal interactions have treated the particle-particle interactions under the assumption that the overlapping electrical double layers equilibrate rapidly and continuously as the particles approach each other. The Gibbs energy for interaction is, therefore, calculated using classical thermodynamics rather than using the framework of irreversible thermodynamics. The implication is that the interacting double layers are always relaxed, an assumption that has received increasing criticism in the modern literature [Lyklema (1980, 1981)]. In fact, this issue has been noted, in its simplest form, in the monograph of Verwey and Overbeek (1948) in terms of the two limiting cases: (i) completely relaxed double layers, and (ii) totally unrelaxed double layers. The first corresponds to constant potentials at the interacting surfaces and the second to constant charge densities. Since the publication of the experimental studies of Frens, Engel and Overbeek (1967) on silver iodide sols (traditionally considered as a model candidate for constant-potential interaction), interest in the dynamics of interfacial electrochemistry has increased sharply, as it was evident that the rate of double layer relaxation was not rapid enough to guarantee constant-potential interaction even in the case of AgI sols [see, also, Lyklema and van Leeuwen (1982)]. This consideration leads to the need for detailed examination of the rate of relaxation of the double layers relative to the rate of particle collision. Several discussions of interaction potentials under various conditions have appeared in the literature [see Rajagopalan and Kim (1981)], and some of the references and the questions they address are summarized at the end. The reason for the interest in these is obvious when one sees the differences in the potential barriers estimated on the basis of the various resulting expressions (see Figure 4.A.1). An excellent treatment of the basic concepts that need examination in this regard is given by Lyklema (1980).

Two areas of current research, without which the discussion of interfacial dynamics will be incomplete, are the structure of the interface itself and the regulation of charges at the interface.

The DLVO theory is worked out mainly for diffuse double layers (with point charges distributed in a 'diffuse' layer governed by the forces among the charges and by the thermal motion); however, the effects of the finite sizes of the ions and the effects of specific adsorption of ions (leading to the Stern layer) cannot be ignored in the interpretation of colloid stability and Schulze-Hardy rule. This aspect of interfacial structure and its relevance to the dynamics of double layers are discussed by Lyklema (1981).

As mentioned earlier, classical DLVO description of interaction is cast in terms of smooth, mean interfacial potentials or charges. An alternative approach to this mean field approximation is to consider an array of well-defined, chemically distinguishable surface groups on the interface. The charges on the approaching surfaces are assumed to ionize as necessary to minimize the total free energy. An analysis based on this regulated interaction is given by Healy, Chan and White (1980). This analysis is a modification of DLVO theory and is composed of many elements of the conventional

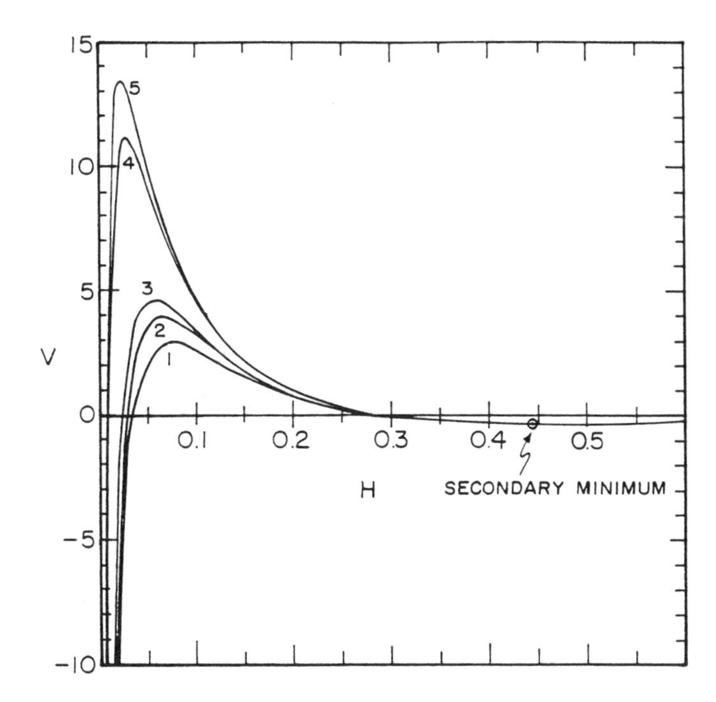

Figure 4.A.1 The effect of change in the mode of double layer interaction on the interaction energy profile for otherwise identical conditions [from Rajagopalan and Kim (1981)].

Curve 1 corresponds to constant potential interaction and Curve 5 to constant charge case. The others are intermediate situations discussed in Section A.3 of Chapter 3. Notice that the magnitude of the barrier determines stability. The interaction energy, V, is in multiples of kT, where k is the Boltzmann constant and T is the absolute temperature of the dispersion. The surface-to-surface distance, H, is in multiples of particle radius.

theory. It retains much of the formalisms of the earlier versions [Ninham and Parsegian (1971); Chan et al. (1975); Chan, Healy and White (1976)]. An alternative formalism is also available in the literature [Ash, Everett and Radke (1973)]. The main conclusions based on this approach are still largely qualitative and can be summarized as follows:

(i) The examples discussed by Healy, Chan, and White (1980) show that the charge regulation concept may be more useful for surfaces with ionizable groups (polymer latex particles, biocolloids, etc.) than the conventional mean field approach.

(ii) Surfaces that can regulate charges will in general coagulate at lower salt concentrations and will form ordered structures at higher volume fractions than surfaces that cannot regulate.

(iii) The use of regulated interaction may be more flexible as a model and may provide a better understanding of interaction of double layers than the usual constant-potential or constant-charge approximations.

A.4 Refinements of Equations for Double Layer Interaction Potentials

In addition to the improvements to the theory of electrical double layer interactions discussed in Section A.3 above, other refinements and sophistications have been added in recent years to the solution of the Poisson-Boltzmann equation and to the attendant approximations. These *mathematical* refinements, which have proceeded in a parallel development to the *physical* refinements cited above, include some of the modifications to the interaction energy expressions mentioned in the context of interfacial dynamics. To these one must add the investigations on the higher-order corrections to the Hogg-Healy-Fuerstenau class of formulas at moderate potentials [Ohshima, Healy and White (1982a,b)] and double layer curvature corrections [Ohshima et al. (1983)]. A two-dimensional version of the Poisson-Boltzmann equation has also been solved, by Barouch et al. (1978), and this leads to much lower barriers in the interaction energy than those predicted by original

Hogg-Healy-Fuerstenau (1966) solution. Matijević, Kuo and Kolny (1981) have subsequently reported that these lower potential barriers are more in line with experimental observations of deposition data than the Hogg-Healy-Fuerstenau values.

A.5 The Role of Structural Forces

In recent years, improved experimental procedures for measuring interaction forces between smooth surfaces have brought to light the importance of structural or hydration forces at small distances of separation ($\leqslant 30$ Å). Continuum theories break down at these separations and are valid only for the long-range tails of the interaction forces. The structural forces can be attractive or repulsive, oscillatory or monotonic, and arise due to surface-induced solvent structure and liquid-induced interface structure [see Ninham (1980, 1982)]. These forces and their experimental determination are discussed briefly in Chapter 3. It seems intuitively reasonable that these forces will have significant influence on stability of dispersions, foams, etc., and on adsorption and desorption of macroions and particles on surfaces; however, this branch of colloid science is still in its infancy.

A.6 Repeptization

No discussion of electrostatic stabilization will be complete without a reference to redispersion or repeptization of flocculated sols. Repeptization has been demonstrated and measured quantitatively by Frens and Overbeek (1971). Both thermodynamic and kinetic aspects are important for understanding repeptization (see Figures 4.A.2 and 3). The types of interaction energy diagrams that allow redispersions of flocculated sols have been investigated by Frens and Overbeek (1972), Overbeek (1977) and Frens (1978).

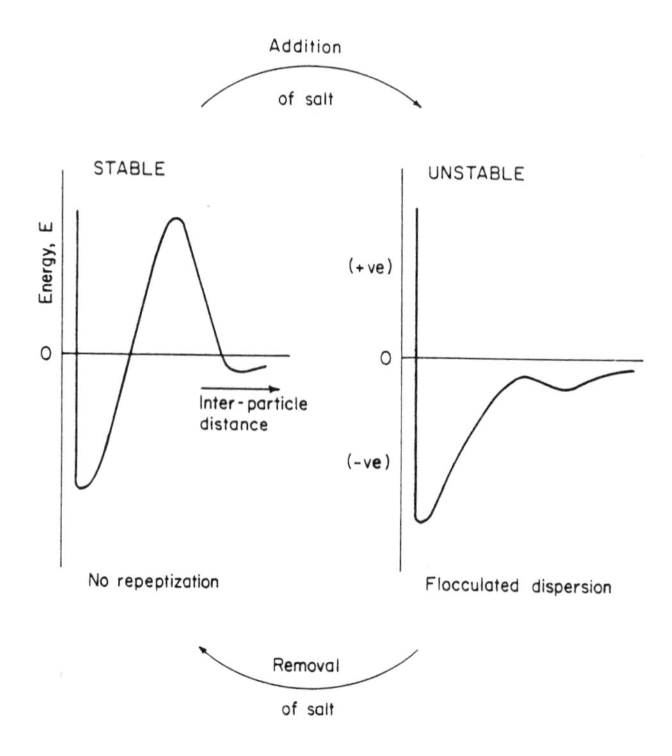

Figure 4.A.2 Effect of addition and removal of salts on coagulation and repeptization.

Addition of salt lowers the barrier and the dispersion flocculates. Subsequent removal of salt restores the barrier, and repeptization is prevented. However, whether repeptization is favored or not depends on the type of interaction energy profile that results on removal of salt; see Figure 4.A.3.

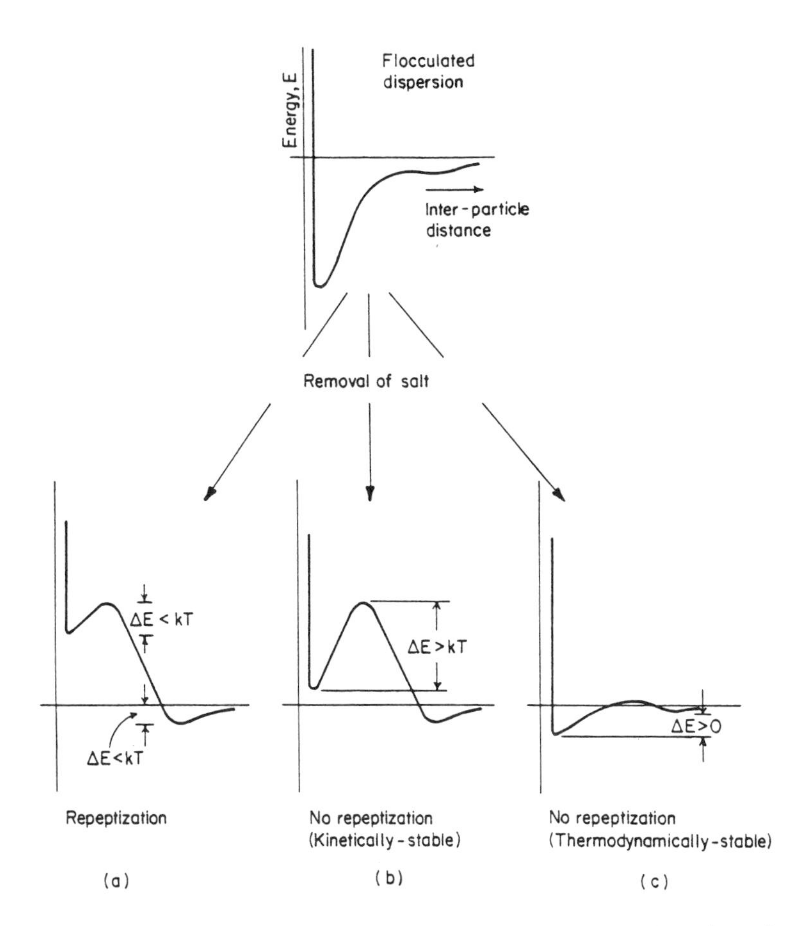

Figure 4.A.3 Understanding repeptization requires both thermodynamic and kinetic considerations.

The top Figure corresponds to an interaction energy profile that permits flocculation. On removal of salt, if the profile shown in Figure (a) in the bottom is obtained, repeptization occurs spontaneously. If, on the other hand, profiles shown in (b) or (c) occur, repeptization will not occur [for kinetic reasons in the case of (b), since the barrier is large; for thermodynamic reasons in the case of (c), since the free energy will increase if repeptization occurs]. The qualitative description above is based on the assumption that the actual potential of mean force is sufficiently accurately given by the average pair interaction potential.

The approaches are heuristic and empirical and are based on approximations to the shape of interaction energy profiles at or near the primary minimum. As noted by Frens (1978), these approaches again draw one's attention to the importance of the potential across the diffuse part of the double layer and hence to the interfacial structure (see Section A.3 above). Although repeptization and the analogous problem of removal of particles adsorbed on surfaces (see Chapter 5) assume great importance in practical and fundamental aspects of colloidal phenomena, their analyses largely rest on the success of investigations on the problems discussed in the previous sections.

A.7 Effects of Bulk Motion on Stability

The effects of shear and other fluid dynamic aspects of bulk motion are certainly extremely relevant and have begun to attract increasing attention in the literature. These are important enough to form a distinct area of study and, hence, are discussed separately in Section C below.

In closing this section, the following general observations are appropriate.

Although electrostatic stability has been one of the oldest and most important issues in colloid science, the number of studies that are sufficiently systematic and detailed to be useful is limited. The reason is that the basic physical and chemical issues are so complex that the description of macroscopic parameters, microstructural details and their interaction is difficult both theoretically and experimentally. Yet, the progress has been sufficiently steady and encouraging. In particular, the DLVO theory has been generally successful under the following restrictions [see Lyklema (1981)]:

(a) The dispersions must be dilute so that the potential of mean force may be replaced by the average pair potential. [The deviation of the potential of mean force from the average pair potential is caused by the onset of many-body interactions, and this, in turn, may necessitate modifications in the analysis of electrostatic interactions; see Chapter 6 of this Report; see, also, Castillo, Rajagopalan and Hirtzel (1984).]

(b) No other interaction forces besides van der Waals attraction and double layer repulsion are important;

(c) The geometry of the particles is relatively simple (e.g., smooth spheres); and

(d) The double layers are purely diffuse.

The state of the research so far has been summarized above; the directions for research will be outlined at the end of this Chapter.

B. Polymer Stabilization

From the industrial point-of-view, steric stabilization has many advantages that are lacking in the case of electrostatic stabilization. Perhaps the most important of these is that steric stabilization is equally effective in both aqueous and non-aqueous dispersion media. In addition, sterically-stabilized dispersions display relatively low viscosities, and their stability is relatively insensitive to the electrolyte concentrations. Good freeze-thaw stability is also attributed to these dispersions. These advantages perhaps explain why natural or synthetic steric-stabilizers are exploited in a wide range of industrial products such as paints, inks, pharmaceutical and food emulsions and are favored in treatment and recovery processes such as tertiary oil recovery, water treatment and soil stability. Stability in many biological systems (e.g., milk and blood) is frequently sterically-induced.

One certainly pays a price for these advantages, and this takes the form of increased complexity in understanding the qualitative and quantitative action of polymeric stabilizers in promoting stability. The theory and practice of polymeric stabilization are at least as complex as electrostatic stability, for these require the full power of the thermodynamics of polymer solutions for a full understanding. Although a complete theoretical understanding of polymeric stabilization is a lofty goal that is certainly beyond our reach now, that has not discouraged the wide-spread use of this technique, as mentioned above, and, in fact, the industrial use of polymeric stabilizers seems to have developed into a fine, empirical art [see the monograph by Sato and Ruch (1980)].

The nonionic polymers used for imparting stability can also work in another manner that has not been mentioned so far. This latter mode of operation is known as depletion stabilization and is described separately in Section B.3 below.

In addition to the monograph cited above, excellent overviews of polymer stabilization are available in the literature by a major contributor to this area [Napper (1977, 1982, 1983)], and these may be consulted for lengthier and deeper discussions of the technical details. Recent reviews of polymer adsorption at interfaces and its effect on the stability of dispersed systems are also available in Vincent and Whittington (1982) and Goddard and Vincent (1984). Sato and Ruch (1980), in particular, is a good source for the historical development of ideas in this area and has a good collection of free energy expressions needed for the computation of the extent of steric stability under various conditions. A qualitative (and somewhat exaggerated) picture of steric action is depicted in Figure 4.B.1. Despite the earlier assertion that this area is still largely an art, one must concede that considerable progress has been made in the last two decades in certain crucial problems on polymer behavior in the bulk and at the interfaces. Again, for reviewing these advances in a tractable manner, it is convenient to divide them into three categories:

1. Steric Stabilization: Phenomenological Aspects,
2. Steric Stabilization: Theoretical Aspects, and
3. Depletion Stabilization.

These are taken up below.

B.1 Steric Stabilization: Phenomenological Aspects

Under this general heading we shall consider the current understanding of factors such as critical flocculation point, particle concentration dependence and classification of sterically-stabilized dispersions. Empirical

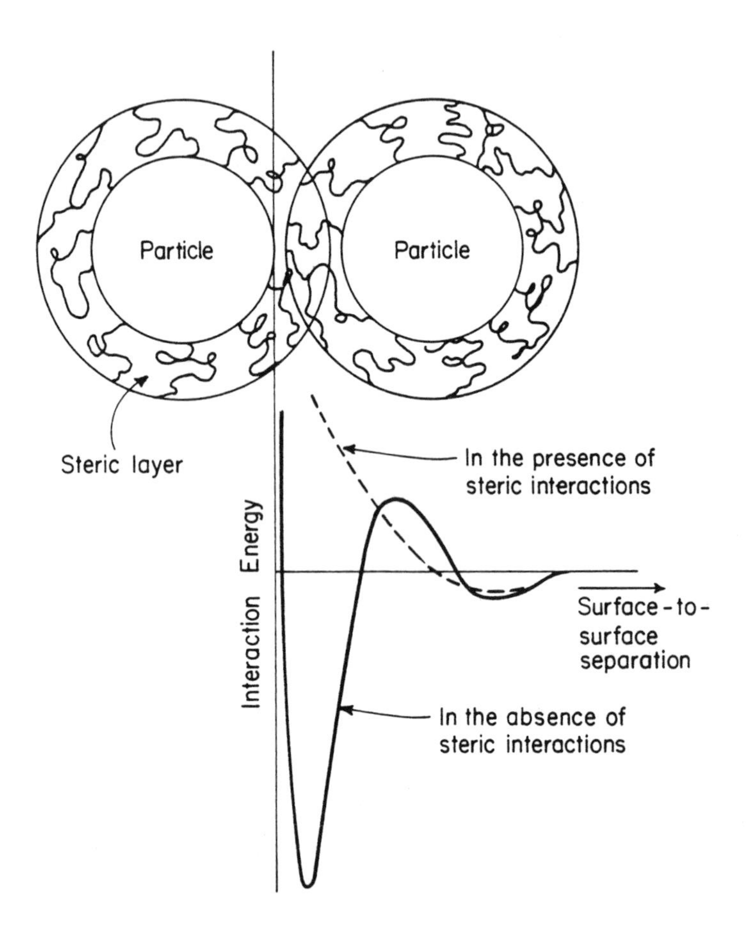

Figure 4.B.1 A qualitative picture of steric interaction.

observations have shown that amphipathic polymers perform best in both aqueous and non-aqueous media. These consist of nominally insoluble anchor polymers attached to nominally soluble stabilizers. The role of anchor polymers in preventing desorption of stabilizers under stress during collisions with other particles and the extent of coverage needed for optimizing stability are well-understood [see Barrett (1975)]. Homopolymers, on the other hand, are claimed to be poorer in performance because of the conflicting requirements that the dispersion medium be a poor solvent to ensure strong adsorption of the stabilizer on the particles but a good solvent to impart effective stabilization. Instability is induced by decreasing the solubility of the stabilizer in the dispersion medium, and this is accomplished by varying temperature or pressure or by the addition of a miscible non-solvent for the stabilizer to the medium. The transformation from long-term stability to flocculation is sudden under these inducements, and the value of temperature, pressure or volume at which this occurs is called the critical flocculation point (cfp). The relation between the cfp and the theta-point [see Sato and Ruch (1980)] for the stabilizer in a solution of the dispersion medium has been studied extensively by Croucher and Hair (1978, 1979, 1980a,b) and Napper and co-workers [see Napper (1982)]. The dependence of cfp on the particle number concentration in the dispersion has also been examined experimentally; see Napper's review for references and additional information. The classification of sterically-stabilized dispersions is based on the sign of the change in Gibbs free energy of close approach of the particles. Both Sato and Ruch (1980) and Napper (1982) present various combinations of entropic and enthalpic changes and discuss the implication of these changes to the classification of the resulting stability or instability. In general, the following observations may be made concerning the phenomenological aspects:

(i) Thermodynamic factors limit the stability of sterically-stabilized dispersions (if the anchoring is adequate and surface coverage is complete). It is known empirically that the nature and size of the particles generally have no influence on this, unless the particles are large with thin steric layers. (In that case, van der Waals attraction, which is otherwise unimportant, may promote flocculation.)

(ii) There are some disagreements in the literature regarding the dependence of cfp on particle concentration. To what extent the loss of configurational entropy of the particles contributes to the above dependence may require additional attention.

(iii) In many sterically-stabilized dispersions, van der Waals forces are too weak to cause flocculation, in marked contrast to electrostatically-stabilized dispersions.

B.2 Steric Stabilization: Theoretical Aspects

Qualitative and quantitative attempts to understand the interaction energies in steric stabilization rely on recent (and ongoing) research on polymer solution thermodynamics and hence are open to differing interpretations. Many of the components of these attempts are drawn from the Prigogine-Flory free volume theory (see the papers of Croucher and Hair cited earlier), but in order to bring into focus the research directions explored in this connection and their current status, we may use Napper's summary of three effects of free volume theory identified to be operative in steric stabilization. These are

1. Combinatorial effects arising from the mixing of polymer segments with the molecules of the dispersion medium. These can be calculated using the Flory-Huggins expression for the entropy of mixing. These effects always oppose flocculation,

2. Free volume effects arising from the dissimilarities in the free volumes of the solvent molecules and the articulated macromolecules. The corresponding free energy may have both entropic and enthalpic contributions, and

3. Contact dissimilarity effects, which again may lead to entropic or enthalpic contributions to free energy.

The relevance and nature of these effects in both aqueous and non-aqueous media are discussed in Napper (1982). Within this general framework, quantitative ab initio predictions of interaction energies depend on, as well-known, the resolution of two general problems, namely,

a. prediction of polymer conformation at the interface, and,

b. the thermodynamics of interactions between two steric barriers.

Flory (1969) has descriptions of the statistical mechanical models of the conformational properties of polymeric chains, and attempts to model the conformational behavior of polymers in solvents that are better than θ-solvents are also available in the literature [Wall et al. (1978); Guttman et al. (1978)]. The presence of an interface, obviously, complicates the problem because of potential specific interactions of the polymer with the interface. Some special cases in this category have been studied using Monte Carlo procedures by Napper and co-workers [Feigin and Napper (1979); see also Napper (1982)]. Because of the deficiencies in the practical application of the Flory-Huggins theory, Evans and Napper (1977) have explored the use of semi-empirical approaches for exploiting the above theory (by determining the appropriate interaction parameters through experiments and then using the Flory-Huggins expressions).

On the second problem of interpreting interactions between steric layers on close contact, the approach has been to divide the free energy on close contact into a mixing contribution and an elastic contribution. These do not have to be computed separately, but Sato and Ruch (1980) discuss the theories available to estimate the individual contributions. [Sato and Ruch (1980) use the terms 'osmotic' and 'volume restriction' for the above two effects, respectively. The use of the term 'osmotic', instead of 'mixing', has been criticized by Napper (1982) as inappropriate, since, from a fundamental thermodynamic point-of-view, almost all types of colloid stabilization mechanisms can be brought under the label 'osmotic'.]

All theories of steric stabilization use the above framework, and, as mentioned earlier, attempt either a completely predictive approach starting from the description of polymer conformation at the interface [see, for example, Scheutjens and Fleer (1979, 1980); de Gennes (1979)] or use a semi-empirical approach cited earlier.

The above discussion provides the background against which the following summary of research trends must be viewed:

(i) The general qualitative aspects of the thermodynamical arguments used for estimating free energy changes are on a better footing for non-aqueous systems than for aqueous systems. For instance, hydrophobic interactions may affect the entropic effects in biopolymers. In view of these, free energy estimates for close approach in the case of aqueous polymer solutions are open to speculation.

(ii) The problem of predicting conformational behavior of polymers is still an active area of research and is clearly very difficult. The conformational properties of polymers in solvents of interest in steric stabilization are receiving increasing attention, but these and the associated problems are too complex to permit rigorous theoretical analyses at this time.

(iii) The more difficult problem of studying conformational properties of polymers at interfaces is beginning to receive attention.

(iv) The more pragmatic approach of calculating free energy contributions using some experimental information on factors such as segment density distribution functions (using neutron scattering) seems to be a workable idea. In fact, such approaches have led to important insights concerning the possibility of complete thermodynamic stability in the case of steric stabilization.

Finally, a few comments concerning some parallel developments that provide some support for the above theoretical approaches are in order. Direct measurements of interaction energies between polymer-coated mica sheets [Klein (1980)] and indirect analyses through osmotic pressure measurements on sterically-stabilized polymer latices [Cairns et al. (1976)] have been done. The former of these seems to confirm the general shape of the interaction energy curves for the cases studied. Monte Carlo results [Cairns, van Megen and Ottewill (1981)] and analytical procedures [Evans and Napper (1978)] for computing osmotic pressures are slowly being introduced for checking the experimental values, and these form a fruitful area of enquiry. A few other experimental studies, on the phenomenological aspects concerning enhanced steric stabilization and steric stabilization in polymer melts, are reviewed in Napper (1982).

B.3 Depletion Stabilization

Free polymers at low concentrations can add stability to or destabilize dispersions. Flocculation is favored thermodynamically when the polymer molecules are excluded from the interparticle region, since in such cases mixing of almost pure solvent (from the interparticle regions) with the bulk solution decreases the free energy of the system; this promotes contacts between the particles. This phenomenon is known as depletion flocculation. The opposite effect is also possible, when rejection of polymer molecules from the interparticle region is not favored thermodynamically, as in the case of good solvents. This effect is, of course, known as depletion stabilization. The existence of such an effective potential barrier in interaction energy makes depletion stabilization analogous to electrostatic stabilization; that is, stability in this sense is a thermodynamic metastability (in contrast to steric stabilization, which, as mentioned earlier, can be thermodynamically stable).

The status of research on these can be summarized as follows:

(i) A general theoretical understanding of depletion stabilization exists presently [see Napper (1982)] for the case where the particle surfaces are free of polymers.

(ii) Presence of some steric layer on the particles complicates this analysis since in this case the steric effects and depletion effects are non-linearly coupled; i.e., the interaction energies are not additive. This problem needs further study.

(iii) The conflicting actions of steric effects and depletion effects can mean that flocculation is possible in a restricted interval of polymer concentrations; that is, at very low polymer concentrations steric effects may dominate. At large concentrations, depletion stabilization will take over for reasons mentioned above. In the intermediate values, depletion flocculation will come into effect. These phenomenological observations are only empirically understood at present.

A schematic diagram of the effect of an idealized polymer [on the various aspects of steric stabilization (and destabilization) discussed in this Section] is presented in Figure 4.B.2. This representation is due to Fleer [see Napper (1982)] and summarizes the discussions presented in this Section. A recent, but limited, discussion of some outstanding problems on steric stabilization is available in Osmond (1983). A more extensive list of research needs is presented at the end of this Chapter.

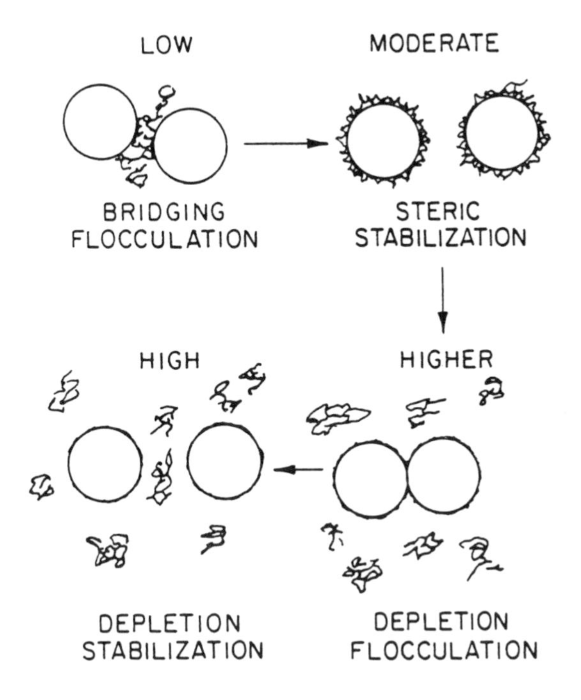

Figure 4.B.2 Schematic representation of the effects of polymers on stability [from Napper (1982)].

C. Bulk Motion and Colloid Stability

The importance of bulk motion in promoting coagulation or redispersion has been recognized at least since the time of the appearance of Smoluchowski's 1917 paper on coagulation in laminar shear fields [Smoluchowski (1917)], although research on this aspect of colloidal stability has always lagged behind that on electrostatic and steric stabilization. The major reason for this can be traced to the fact that the birthplace of the classical studies on coagulation was the domain of the experimental physical chemist. In addition, chemical aspects of the phenomenon were the first to attract attention, were for a long time considered to be the most immediate for a proper understanding of coagulation, and have been, even to this day, the most dominating. Notwithstanding the early and pioneering contributions of the gifted physicist Smoluchowski, the physics of coagulation, to which the influence of fluid dynamics on stability rightly belongs, has so far been slow to develop.

A contributing factor to this has been the difficulty of describing, with sufficient rigor, the fluid dynamics of multiparticle systems, an area labeled microhydrodynamics by Batchelor (1976a). There have been significant developments in the dynamics of suspensions in liquids in recent years [Herczynski and Pienkowska (1980)], and outstanding treatments of suspension behavior at both dilute and moderate concentrations, purely from the perspectives of fluid dynamics, are available in the literature [see, for example, Happel and Brenner (1973) and Brenner (1974)]. We shall, however, restrict our focus here to specifically those studies that pertain to colloid stability or show potential for extensions in that direction. Additional information on the former may be found in the references cited above and in the recent report on slurry and suspension transport prepared by Turian (1982) for the U. S.

National Science Foundation.

To be precise, microhydrodynamics has been generally taken to mean low Reynolds number (i.e., non-inertial) hydrodynamics -- a restriction readily satisfied by the conditions that normally prevail in flows around particles of colloidal dimensions. The increase in research activities in the last decade on flow-induced coagulation (and, to a lesser extent, redispersion) has been spurred largely by the increase in our understanding of the hydrodynamic field around spherical particles taken in pairs. In what follows, these recent advances are reviewed briefly. Each subsection addresses a major direction in the area. It is, however, important to set forth, at the outset, the following limitations of the current activities:

(i) The bulk of the research activities is concerned with interactions between only two isolated particles;

(ii) Most of the work is concerned with homocoagulation, i.e., collisions between particles (more correctly, spheres) of same size, except where noted;

(iii) Inclusion of colloidal interactions has been minimal, although the potential for improvements in this regard is substantial.

Within these general constraints, the promising research directions in this area are the following.

C.1 Influence of Viscous Drag on Collision Rate

The decrease in collision rates between two particles caused by increased viscous resistance as the two particles approach each other along the axis connecting their centers was discussed in Section A.2 above, in the context of its influence on electrostatic stability. This decrease in collision rates is well-understood for the case of particle motion in stagnant Newtonian fluids, and the discussion in Section A.2 applies to the general case of spheres of arbitrary radii. Computed values for collision rates in the presence of van der Waals attraction are available in Kim and Rajagopalan (1982); see Figure 4.C.1. The situation in which there is a bulk fluid motion, which causes the particles to approach each other along the axis, must be viewed in the general context of arbitrary flow fields and is discussed in the following sections.

C.2 Influence of Shear on Coagulation

The earliest work in recent years that improved upon the original contribution of Smoluchowski on shear-induced coagulation is that of Curtis and Hocking (1970), who considered the effect of laminar shear field on the collision rates of two, equal-sized, uncharged spheres interacting through van der Waals attraction. Although Curtis and Hocking used the computed collision rates to interpret their experiments and to estimate reasonable values for the effective Hamaker constants for their dispersions, the solution for the fluid field used in their calculations [from Brenner (1961)] is not appropriate for small separations. Brenner's series solution converges very slowly for small separations and ultimately diverges when the spheres make contact. This problem was rectified later by Cooley and O'Neill (1969) using matched asymptotic expansions to obtain the resistance experienced by the spheres. Subsequently, Batchelor and Green (1972) presented improved hydrodynamic solutions for binary interactions between neutrally-buoyant spheres (of equal size) using the Cooley-O'Neill results. These have been used by van de Ven and Mason (1976a,b,c) and Zeichner and Schowalter (1977) to improve the calculations of Curtis and Hocking (1970). The papers by van de Ven and Mason incorporate an additional improvement over that of Curtis and Hocking through the inclusion of DLVO effects; however, these calculations are restricted to laminar shear flow. Zeichner and Schowalter (1977) is more general and

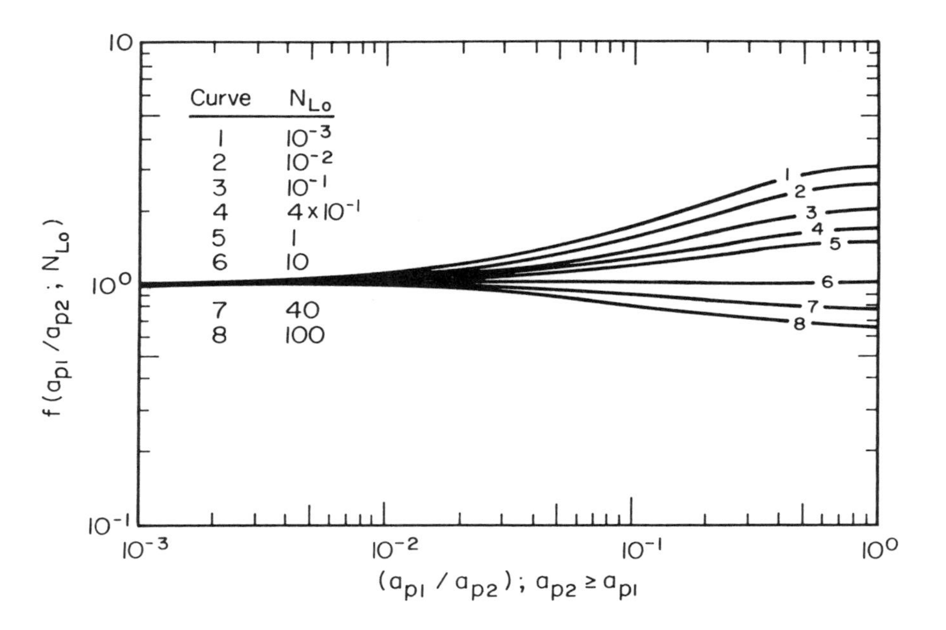

Figure 4.C.1 Influence of hydrodynamic retardation on the collision rates (or, equivalently, stability ratios); from Kim and Rajagopalan (1982).

At low values of attraction (i.e., dimensionless Hamaker constant, N_{Lo}, much less than 1), the stability ratio f increases because of hydrodynamic retardation. For collision between a small particle and a large object [i.e., (a_{p1}/a_{p2}) much less than 1], the effects of attractive forces and hydrodynamic retardation are negligible and cancel each other.

considers non-equatorial trajectories with uniaxial extensional as well as laminar shear flows. Both contributions have considered the vorticity effects on capture efficiency that can be anticipated from the work of Kao and Mason (1975). Additional information on the technical aspects can be obtained from the above references and from Schowalter (1982, 1984) and van de Ven (1982). For our purpose here, the following general observations are relevant:

(i) Using trajectory calculations, stability domains (for primary minimum coagulation, secondary minimum coagulation and stability against coagulation) can be constructed for coagulation in bulk flows in the presence of attraction and repulsion [Schowalter (1982)]; see Figure 4.C.2.

(ii) Availability of the Batchelor-Green solution for hydrodynamically-interacting spheres permits comparison of the effects of individual flow fields (e.g., laminar shear, uniaxial extensional, etc.), although the computations that have been made so far are far from complete.

(iii) Trajectory calculations also permit the examination of orbiting trajectories [see Schowalter (1982) and Feke and Schowalter (1983)].

(iv) Despite the reasonable values of Hamaker constants obtained using these techniques, the coagulation rate is sensitive to the Hamaker constant (in the absence of repulsion) and hence there is considerable uncertainty in this procedure for the determination of the Hamaker constants.

(v) When electrostatic repulsion is important, agreement between theory and experiments is less certain.

Additional references of related interest are available in the above publications [see, also, Zeichner and Schowalter (1979)]. Before leaving this section and proceeding to other theoretical and experimental advances in this area, an observation concerning the Batchelor-Green solution for binary, hydrodynamic interactions referred to above will be useful in the context of the discussions presented later in this Report in the material on Concentrated Dispersions. The binary interaction solutions of the type presented in Batchelor and Green (1972) and Lin, Lee and Sather (1970) were motivated

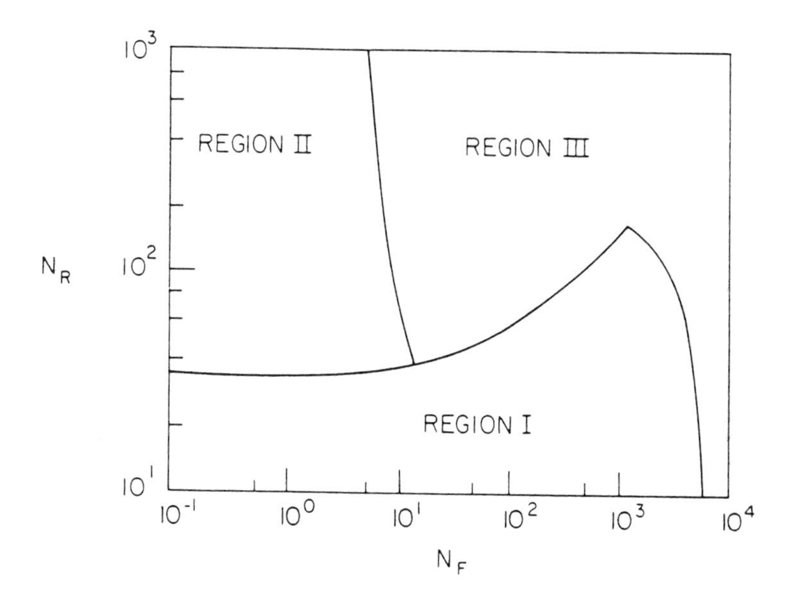

Figure 4.C.2 Stability map for laminar shear flow [from Schowalter (1982)].

Region I corresponds to primary minimum flocculation, Region II to secondary minimum flocculation, and Region III to a stable dispersion. The dimensionless group, N_R, is a repulsion parameter and represents the ratio of repulsion to attraction; N_F is a measure of the hydrodynamic forces relative to attractive forces.

originally by the need to obtain rigorous expressions for the viscosity of dispersions at concentrations away from the limit of infinite dilution; however, the velocity fields and shear fields computed using these solutions can be combined with the assumption of pair-additivity to obtain diffusion coefficients in concentrated dispersions [Venkatesan (1984)]. These will be discussed further later in this Report.

C.3 Brownian Motion and Bulk Flow

The progress in combining Brownian motion effects and bulk flow effects has been slow and frustrating. The simplest assumption that can be made is that these two mechanisms act independently of each other without any mutual influence. This assumption permits simple additive combination of the two effects in the general case and was invoked by Swift and Friedlander almost twenty years ago [Swift and Friedlander (1964)]. The results of their calculations, for Brownian motion combined with laminar shear flow, were in good agreement with their experiments (conducted using a cylinder Couette viscometer) over a fairly large interval of the rate of shear. These calculations and experiments did not consider other colloidal interactions. This status has changed recently, and theoretical and experimental studies, using the generalized thermodynamic formalism of Batchelor (1976b) for the effective Brownian velocity, have been attempted for small and large Peclet numbers [see van de Ven and Mason (1977) for small Peclet numbers and Feke and Schowalter (1983) for large values; the paper of Batchelor (1976b) assumes that the local configuration of the particles remains unaffected by the fluid velocity field. This has been corrected subsequently in Batchelor (1982); see related discussions in Section B of Chapter 7]. Again, the calculations of Feke and Schowalter (1983) have been performed for both uniaxial extensional and laminar shear flows. The probability density (of particle positions) needed for these computations has been calculated using a general form of steady-state convective diffusion equation [see Schowalter (1984) and Feke and Schowalter (1983)]; i.e., the assumption of linear additivity of Brownian motion effects and bulk-flow effects is not necessary and, in fact, is found to be inaccurate. In the case of laminar shear flow, these computations can become very difficult because of the influence of orbiting trajectories on the probability distribution of the positions of the particles. Apart from the above observations, these developments indicate that the general case of coagulation under Brownian motion, bulk flow and colloidal forces is within our reach for further careful scrutiny. In contrast to the above studies on homocoagulation (i.e., coagulation in monodisperse suspensions), studies on heterocoagulation are only now coming under examination [see Adler (1981a,b)] and have not yet yielded any definitive insights.

C.4 Gravity-Induced Instability and Other Topics

Even in the absence of externally-imposed bulk motion, differential sedimentation rates in polydisperse systems can exert a significant influence on stability. In the absence of Brownian motion, trajectory analyses of the type discussed above (and in Chapter 5 in the context of deposition phenomena) can be used to construct stability diagrams [see Melik and Fogler (1984a)]. Under these conditions, depending on the magnitude of physical parameters, dilute spherical sols can flocculate at low or high values of gravitational forces while remaining stable at intermediate values. When Brownian motion is important, stochastic trajectory analyses are possible but it is more profitable to employ appropriate Smoluchowski equations for computing stability ratios. A perturbation solution available for such a case [see Melik and

Fogler (1984b)] shows that it is not always accurate to compute the stability ratios in the presence of Brownian motion by simply summing the individual stability ratios due to diffusion and gravity. Numerical results are presented in Melik and Fogler (1984b). Some additional disucssion on interaction effects on sedimentation rates is presented in Chapter 7, Section B, of this Report.

In addition to the directions reported above, it should be noted that other experimental and theoretical techniques also are becoming feasible. Among these are experimental studies of trajectories of spherical particles [see, for instance, Takamura, Goldsmith and Mason (1981)] and fluorescence techniques for studying coagulation of colloids [Cummins et al. (1983)] which show promise. Theoretical and experimental studies on floc dynamics are also being initiated [see, for example, Hunter (1982)]; these latter studies are also relevant in the rheology of dispersions. Since most of these are concerned primarily with the fluid dynamic aspects of the problems, these will not be discussed further here.

Research Needs

Some of the basic problems concerned with electrical double layers and their structure (particularly in asymmetric electrolytes) have been discussed in Chapter 3. These problems are, of course, important in the context of electrostatic stability of colloidal systems (and deposition phenomena discussed in Chapter 5). In addition to these, interaction of overlapping double layers complicates the analysis of stability phenomena further.

- The dynamic aspects of interfacial chemistry and interacting double layers play a prominent role and need further attention. The charge regulation concept discussed in Section A.3 seems promising.

- Measurement and analysis of dielectric changes as a function of the frequency of an applied electric field have also been suggested for studying relaxation of electrical double layers [see Lyklema (1980) cited in Section A].

- The role of short-range structural forces is only now beginning to receive attention. In addition, attempts to measure directly the interaction forces between surfaces immersed in liquids may be extended to understand short-range forces in the presence of electrolytes, at least qualitatively.

- If equilibrium structure factor measurements in strongly-interacting systems can be used to provide information on effective pair potentials (see Chapter 6), modern advances in radiation scattering techniques may be taken advantage of to investigate double layer interactions.

- The present state of understanding of the above problems in non-aqueous media and in biological systems is considerably behind what has been achieved in the case of aqueous systems.

- Factors such as surface roughness and chemical and charge heterogeneities on the surfaces have been recognized to be very important for a long time. Good ways of accounting for them are of course needed, but these problems are likely to remain unsolved for some time.

- Experimental work on the kinetics and the thermodynamics of repeptization will be useful.

- Coagulation of non-spherical particles has received very little attention. Since methods to produce particles of various shapes are becoming available [see Chapter 2 (Figure 2.3) and Chapter 5,

Section B.2], some preliminary work in this direction is within reach [see Matijević and Kitazawa (1983) in the reference list for Section A].

Solutions of outstanding problems in steric stabilization require solutions for many unsolved problems on the behavior of polymers at interfaces. The two central problems are polymer conformation at interfaces and the thermodynamics of interactions between steric barriers.

- The configurational properties of polymers in solvents that are better than θ-solvents are still inaccessible. Additional work, such as the Monte Carlo studies mentioned in Section B.2, will be useful.
- Experimental techniques such as neutron scattering can be used to obtain segment density distributions on particle surfaces. Additional work in this area is needed.
- A combination of experimental measurement of equilibrium properties and theoretical study of these (e.g., by using statistical mechanical approaches) may be useful in understanding the thermodynamics of interaction between steric barriers.
- Direct measurements of steric forces using polymer coated mica sheets have led to some understanding of interactions between steric barriers. Additional work along this line is needed (see Section B.2).
- The presence of steric layers on particles complicates the analysis of depletion stabilization since the adsorbed polymers tend to counteract the effects of depletion zones near the particle surfaces. The potential energies due to depletion effects and steric effects are not necessarily additive. This problem needs further study.
- The lateral interactions of adsorbed polymer chains have been ignored for a long time and require reconsideration [see Osmond (1983) and Napper (1982) cited in Section B].

The combined effects of hydrodynamics and colloidal interactions require more attention both from theoretical and from experimental sides.

- In addition to further theoretical work on the combined effects of colloidal forces and fluid flow, simultaneous coagulation in primary and secondary minima (in the presence of flow) requires attention. The need to include Brownian diffusion is unavoidable. Gravity-induced flocculation can also be very significant. Perhaps, a computer simulation approach is the easiest possible currently.

- Steric effects on flow-induced coagulation also need attention. Initial attempts in this direction can be made by treating the particles as spheres with an appropriate effective diameter and using suitable steric potentials.

- Experimental studies on slow coagulation in well-defined flow fields are needed (see Section C.4).

- Theoretical work on heterocoagulation in dilute systems (e.g., interaction between two unequal spheres) is within reach.

- Floc formation and rupture in flow fields are important and will be of use in understanding rheology of flocculated dispersions. This is, however, a formidable problem that will have to await results of simpler cases.

REFERENCES: SECTION A. ELECTROSTATIC STABILIZATION

Ash, S. G., Everett, D. H. and Radke, C., J. Chem. Soc. Faraday Trans. II 69, 1256 (1973).

Barouch, E., Matijević, E., Ring, T. A. and Finlan, J. M., J. Colloid Interface Sci. 67, 1 (1978).

Castillo, C. A., Rajagopalan, R. and Hirtzel, C. S., Rev. in Chem. Eng., in press (1984).

Chan, D., Healy, T. W., Perram, J. W. and White, L. R., J. Chem. Soc. Faraday Trans. I 71, 1046 (1975).

Chan, D., Healy, T. W. and White, L. R., J. Chem. Soc. Faraday Trans. I 72, 2844 (1976).

Derjaguin, B. V., Trans. Faraday Soc. 36, 203 (1940a).

Derjaguin, B. V., Trans. Faraday Soc. 36, 730 (1940b).

Derjaguin, B. V., Discuss. Faraday Soc. 42, 317 (1966).

Derjaguin, B. V. and Landau, L., Acta Physicochim. URSS 14, 633 (1941a).

Derjaguin, B. V. and Landau, L., J. Expt. Theor. Phys. 11, 802 (1941b).

Derjaguin, B. V. and Landau, L., J. Expt. Theor. Phys. 15, 662 (1945).

Derjaguin, B. V. and Muller, V. M., Dokl. Akad. Nauk. SSSR 176, 738 (1967).

Deutch, J. M. and Felderhof, B. U., J. Chem. Phys. 59, 1669 (1973).

Frens, G., Faraday Discuss. Chem. Soc. 65, 146 (1978).

Frens, G., Engel, D. J. C. and Overbeek, J. Th. G., Trans. Faraday Soc. 63, 418 (1967).

Frens, G. and Overbeek, J. Th. G., J. Colloid Interface Sci. 36, 286 (1971).

Frens, G. and Overbeek, J. Th. G., J. Colloid Interface Sci. 38, 376 (1972).

Fuchs, N., Z. Phys. 89, 736 (1934).

Goodwin, J. W. (ed.), Colloidal Dispersions, The Royal Soc. Chem., London, U. K., 1982.

Hardy, W. B., Proc. Roy. Soc. London 66, 110 (1900a).

Hardy, W. B., Z. Phys. Chem. 33, 385 (1900b).

Healy, T. W., Chan, D. and White, L. R., Pure Appl. Chem. 52, 1207 (1980).

Hiemenz, P. C., Principles of Colloid and Surface Chemistry, Marcel Dekker, New York, NY, 1977.

Hogg, R., Healy, T. W. and Fuerstenau, D. W., Trans. Faraday Soc. 62, 1638 (1966).

Honig, E. P., Roebersen, G. J. and Wiersema, P. H., J. Colloid Interface Sci. 36, 97 (1971).

Kerker, M., Zettlemoyer, A. C. and Rowell, R. L. (eds.), Colloid and Interface Science, Vol. 1, Academic Press, New York, NY, 1977.

Lichtenbelt, J. W. Th., Ras, H. J. M. C. and Wiersema, P. H., J. Colloid Interface Sci. 46, 522 (1974).

Lichtenbelt, J. W. Th., Pathmamanoharan, C. and Wiersema, P. H., J. Colloid Interface Sci. 49, 281 (1974).

Lyklema, J., Pure Appl. Chem. 52, 1221 (1980).

Lyklema, J., Pure Appl. Chem. 53, 2199 (1981).

Lyklema, J. and van Leeuwen, H. P., Adv. Colloid Interface Sci. 16, 127 (1982).

Matijević, E. and Kitazawa, Y., Colloid Polym. Sci. 261, 527 (1983).

Matijević, E., Kuo, R. J. and Kolny, H., J. Colloid Interface Sci. 80, 94 (1981).

Ninham, B. W., J. Phys. Chem. 84, 1423 (1980).

Ninham, B. W., Adv. Colloid Interface Sci. 16, 3 (1982).

Ninham, B. W. and Parsegian, V. A., J. Theor. Biol. 31, 405 (1971).

Ohshima, H., Chan, D. Y. C., Healy, T. W. and White, L. R., J. Colloid Interface Sci. 92, 232 (1983).

Ohshima, H., Healy, T. W. and White, L. R., J. Colloid Interface Sci. 89, 484 (1982a).

Ohshima, H., Healy, T. W. and White, L. R., J. Colloid Interface Sci. 90, 17 (1982b).

Ottewill, R. H., pp. 379-395 in Kerker, Zettlemoyer and Rowell (1977).

Ottewill, R. H. and Shaw, J. N., Discuss. Faraday Soc. 42, 154 (1966).

Overbeek, J. Th. G., pp. 431-445 in Kerker, Zettlemoyer and Rowell (1977).

Overbeek, J. Th. G., pp. 1-22 in Goodwin (1982).

Rajagopalan, R. and Kim, J. S., J. Colloid Interface Sci. 83, 428 (1981).

Schulze, H., J. Prakt. Chem. 25, 431 (1882).

Schulze, H., J. Prakt. Chem. 27, 320 (1883).

Shaw, D. J., Introduction to Colloid and Surface Chemistry, 3rd ed., Butterworths, London, U. K., 1980.

Smoluchowski, M., Phys. Z. 17, 557 (1916a).

Smoluchowski, M., Phys. Z. 17, 585 (1916b).

Smoluchowski, M., Z. Phys. Chem. 92, 129 (1917).

Spielman, L. A., J. Colloid Interface Sci. 33, 562 (1970).

Verwey, E. J. W., Chem. Weekbl. 39, 563 (1942).

Verwey, E. J. W., Philips Res. Rep. 1, 33 (1945).

Verwey, E. J. W. and Overbeek, J. Th. G., Trans. Faraday Soc. 42B, 117 (1946).

Verwey, E. J. W. and Overbeek, J. Th. G., Theory of the Stability of Lyophobic Colloids, Elsevier, Amsterdam, The Netherlands, 1948.

Vold, R. D. and Vold, M. J., Colloid and Interface Chemistry, Addison-Wesley, Reading, Mass., 1983.

REFERENCES: SECTION B. POLYMER STABILIZATION

Barrett, K. E. J., Dispersion Polymerization in Organic Media, Wiley, London, U. K., 1975.

Cairns, R. J. R., Ottewill, R. H., Osmond, D. W. J. and Wagstaff, I., J. Colloid Interface Sci. 54, 45 (1976).

Cairns, R. J. R., van Megen, W. and Ottewill, R. H., J. Colloid Interface Sci. 79, 511 (1981).

Croucher, M. D. and Hair, M. L., Macromolecules 11, 874 (1978).

Croucher, M. D. and Hair, M. L., J. Phys. Chem. 83, 1712 (1979).

Croucher, M. D. and Hair, M. L., pp. 497 - 510 in Fitch (1980a).

Croucher, M. D. and Hair, M. L., Colloids Surfaces 1, 349 (1980b).

de Gennes, P. G., Scaling Concepts in Polymer Physics, Cornell Univ. Press, Ithaca, NY, 1979.

Evans, R. and Napper, D. H., J. Chem. Soc. Faraday Trans. I 73, 1377 (1977).

Evans, R. and Napper, D. H., J. Colloid Interface Sci. 63, 43 (1978).

Feigin, R. I. and Napper, D. H., J. Colloid Interface Sci. 71, 117 (1979).

Fitch, R. M. (ed.), Polymer Colloids II, Plenum, New York, NY, 1980.

Flory, P. J., Statistical Mechanics of Chain Molecules, Wiley-Interscience, New York, NY, 1969.

Goddard, E. D. and Vincent, B. (eds.), Polymer Adsorption and Dispersion Stability, ACS Symposium Series, Vol. 240, American Chemical Society, Washington, D.C., 1984.

Goodwin, J. W. (ed.), Colloidal Dispersions, The Royal Soc. Chem., London, U. K., 1982.

Guttman, A. J., Middlemiss, K. M., Torrie, G. M. and Whittington, S. G., J. Chem. Phys. 69, 5375 (1978).

Klein, J., Nature 288, 248 (1980).

Matijević, E. (ed.), Surface and Colloid Science, Vol. 12, Plenum, New York, NY, 1982.

Napper, D. H., J. Colloid Interface Sci. 58, 390 (1977).

Napper, D. H., pp. 99-128 in Goodwin (1982).

Napper, D. H., Polymeric Stabilization of Colloidal Dispersions, Academic Press, London, U. K., 1983.

Osmond, D. W. J., pp. 369-379 in Poehlein, Ottewill and Goodwin (1983).

Poehlein, G. W., Ottewill, R. H. and Goodwin, J. W. (eds.), Science and Technology of Polymer Colloids: Characterization, Stabilization and Application Properties, Vol. II, Martinus Nijhoff, The Hague, The Netherlands, 1983.

Sato, T. and Ruch, R., Stabilization of Colloidal Dispersions by Polymer Adsorption, Marcel Dekker, New York, NY, 1980.

Scheutjens, J. M. H. M. and Fleer, G. J., J. Phys. Chem. 83, 1619 (1979).

Scheutjens, J. M. H. M. and Fleer, G. J., J. Phys. Chem. 84, 178 (1980).

Vincent, B. and Whittington, S. G., pp. 1-118 in Matijević (1982).

Wall, F. T., Seitz, W. A., Chin, J. C. and deGennes, P. G., Proc. Natl. Acad. Sci. USA 75, 2069 (1978).

REFERENCES: SECTION C. BULK MOTION AND COLLOID STABILITY

Adler, P. M., J. Colloid Interface Sci. 83, 106 (1981a).

Adler, P. M., J. Colloid Interface Sci. 84, 461 (1981b).

Batchelor, G. K., pp. 33-55 in Koiter (1976a).

Batchelor, G. K., J. Fluid Mech. 74, 1 (1976b).

Batchelor, G. K. and Green, J. T., J. Fluid Mech. 56, 375 (1972).

Brenner, H., Intl. J. Multiphase Flow 1, 195 (1974).

Brenner, H., Chem. Eng. Sci. 16, 242 (1961).

Cooley, M. D. A. and O'Neill, M. E., Mathematika 16, 37 (1969).

Cummins, P. G., Staples, E. J., Thompson, L. G., Smith A. L. and Pope, L., J. Colloid Interface Sci. 92, 189 (1983).

Curtis, A. S. G. and Hocking, L. M., Trans. Faraday Soc. 66, 1381 (1970).

Feke, D. L. and Schowalter, W. R., J. Fluid Mech. 13, 17 (1983).

Happel, J. and Brenner, H., Low Reynolds Number Hydrodynamics, Martinus Nijhoff, The Hague, The Netherlands, 1973.

Herczynski, R. and Pienkowska, I., Ann. Rev. Fluid Mech. 12, 237 (1980).

Hunter, R. J., Adv. Colloid Interface Sci. 17, 197 (1982).

Kao, S. V. and Mason, S. G., Nature 253, 619 (1975).

Kim, J. S. and Rajagopalan, R., Colloids Surfaces 4, 17 (1982).

Koiter, W. T. (ed.), Theoretical and Applied Mechanics, North-Holland, Amsterdam, The Netherlands, 1976a.

Lin, C. J., Lee, K. J. and Sather, N. F., J. Fluid Mech. 43, 35 (1970).

Melik, D. H. and Fogler, H. S., 'Gravity-Induced Flocculation', J. Colloid Interface Sci., in press (1984a).

Melik, D. H. and Fogler, H. S., 'Effect of Gravity on Brownian Flocculation', J. Colloid Interface Sci., in press (1984b).

Schowalter, W. R., Adv. Colloid Interface Sci. 17, 129 (1982).

Schowalter, W. R., Ann. Rev. Fluid Mech. 16, 245 (1984).

Smoluchowski, M., Z. Phys. Chem. 92, 129 (1917).

Swift, D. L. and Friedlander, S. K., J. Colloid Sci. 19, 621 (1964).

Takamura, K., Goldsmith, H. L. and Mason, S. G., J. Colloid Interface Sci. 82, 175 (1981).

Turian, R. M., Slurry and Suspension Transport, Report No. NSF/OIR-82001, National Science Foundation, Washington, D. C., 1982.

van de Ven, T. G. M., Adv. Colloid Interface Sci. 17, 105 (1982).

van de Ven, T. G. M. and Mason, S. G., J. Colloid Interface Sci. 57, 505 (1976a).

van de Ven, T. G. M. and Mason, S. G., J. Colloid Interface Sci. 57, 517 (1976b).

van de Ven, T. G. M. and Mason, S. G., J. Colloid Interface Sci. 57, 535, (1976c).

van de Ven, T. G. M. and Mason, S. G., Colloid Polymer Sci. 255, 794 (1977).

Venkatesan, M., Structure and Dynamics of Colloidal Dispersions, Ph.D. Dissertation, Rensselaer Polytechnic Institute, Troy, NY, 1984.

Zeichner, G. R. and Schowalter, W. R., AIChE J. 23, 243 (1977).

Zeichner, G. R. and Schowalter, W. R., J. Colloid Interface Sci. 71, 237 (1979).

5

Deposition on Substrates

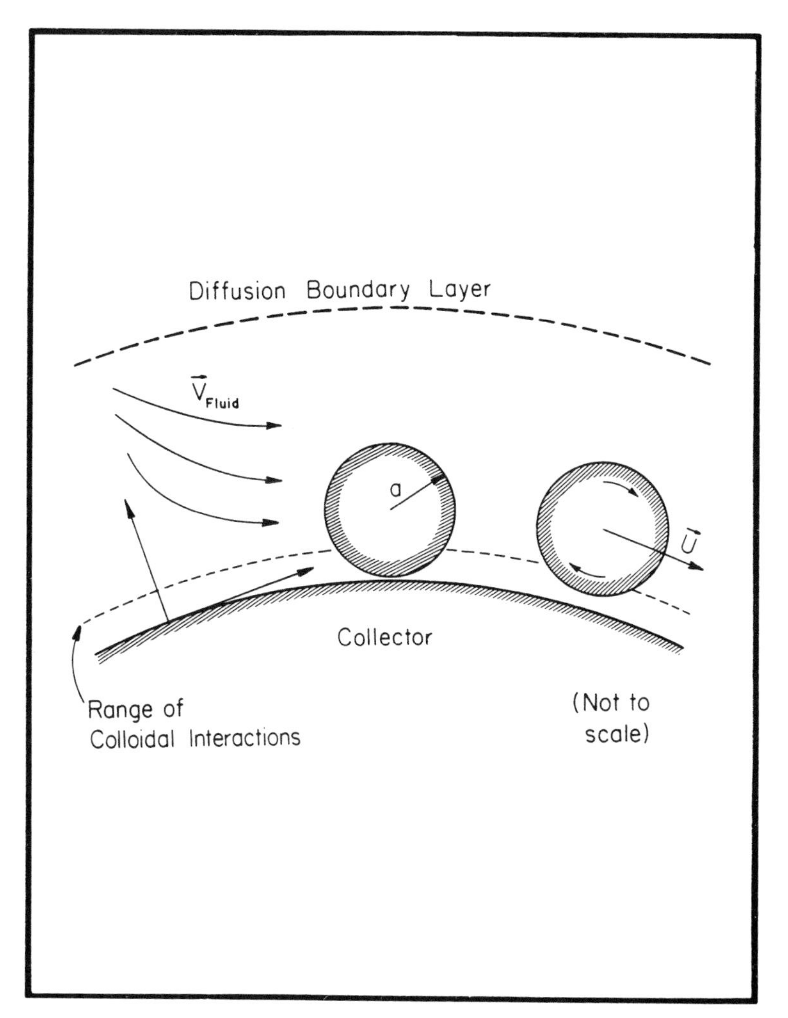

CHAPTER 5

DEPOSITION ON SUBSTRATES

Overview

The phenomena of deposition and resuspension of fine particles or other microscopic species have received much attention during recent years due, in part, to their importance in many natural and industrial processes. These processes span a wide range of engineering applications including, for example, physical separations such as filtration [Rajagopalan and Tien (1977, 1979); Wnek, Gidaspow and Wasan (1975)], biological, chemical and thermal fouling [Somerscales and Knudsen (1981)], virus adsorption [Berkeley et al. (1981); Bitton and Marshall (1980); Marshall (1976); Murray and Parks (1980)], and stability of food and pharmaceutical products. In addition to these, other processes where deposition and adsorption phenomena are relevant include numerous biomedical and physiological processes, e.g., protein adsorption onto cells or other surfaces [e.g., MacRitchie (1978)], immunological analyses such as the 'latex fixation technique' in which antigen is adsorbed onto polystyrene latex particles [Litwin (1977)], and problems related to tissue culture and related studies [e.g., Giaever and Ward (1978); Grinnell (1978)].

Before presenting an overview of the discussion presented in this Chapter, one comment with respect to terminology is needed. The two terms 'deposition' and 'adsorption' are used interchangeably throughout this Chapter for the following reasons. In general, in the case of deposition of colloidal particles, the phenomenon is more appropriately referred to as deposition, rather than adsorption. However, the basic concepts discussed in this Chapter are also applicable to the cases where much smaller species (e.g., proteins and viruses) are involved and for which the term adsorption is usually preferred. Hence, these terms are used interchangeably herein.

The rates of deposition or adsorption of colloidal species on man-made or natural substrates are known to be critically influenced by the physicochemical interactions between the colloidal species and the substrates. Despite the diversity of applications such as those listed above, the underlying physical and chemical phenomena relevant to adsorption or deposition of the particles remain the same. The most important of these interactions are believed to be long-ranged forces such as the London-van der Waals attraction and the coulombic (electrostatic) attraction or repulsion. The shorter-ranged Born repulsive force decreases the depth of the primary minimum where the particles are trapped upon adsorption, and therefore can influence the rates of desorption and resuspension of already adsorbed species. The role of structural forces in deposition, adsorption and resuspension of colloidal particles is not well understood at present and requires further study; currently this is a topic of active research. [For a discussion of the electrokinetic and intermolecular forces involved, see Chapter 3 of this report.]

Attempts to understand the role of physicochemical phenomena on the deposition process are complicated by the sensitivity of the rate to a large number of interdependent parameters and by the practical difficulties in quantifying some of these parameters sufficiently accurately. (Many of these same questions also arise in the study of colloid stability, as discussed in Chapter 4, on Colloidal Stability). Some of the problems and questions which are being studied currently or need attention include the following.

(i) It has been observed experimentally and shown theoretically that the rates of adsorption and desorption are sensitive to the zeta potentials, the chemical types of the ions in the suspension, and the ionic association/dissociation equilibria, among other factors [e.g., Wnek, Gidaspow and Wasan (1975); and Sasaki, Matijević and Barouch (1980)].

(ii) The surfaces of the substrates and of the particles are generally heterogeneous and have localized charges, some of which may be favorable for adsorption, and some unfavorable. This must be taken into account in the interpretation of deposition data.

(iii) The colloidal particles in the suspensions seldom have a single, well-defined electrophoretic mobility. Experimental studies available in the literature are based on observed average mobilities of the particles; however, analyses of data using the average mobility alone can lead to misleading and erroneous conclusions about the adsorption behavior. This is illustrated by Figure 5.1, which shows experimental data on the distribution of particle zeta potential in a 'model' colloid [Rajagopalan and Chu (1982)]. As a result of such distributions, experiments in which the average surface potentials indicate conditions unfavorable for adsorption frequently can, in fact, result in trace amounts of adsorption which, in turn, can show a difference of many orders of magnitude between the observed rates and those predicted theoretically.

(iv) As shown recently by Matijević, Kuo and Kolny (1981), even trace amounts of chemical impurities in the suspension (in concentrations less than the detectability limits of the most sensitive techniques) can sometimes significantly affect the deposition and resuspension processes.

(v) The rate of adsorption also depends on the electrochemical conditions that exist on the interacting surfaces; that is, whether the interacting surfaces keep their potentials constant or their charge densities constant [see Lyklema (1980); see also Sections A.3 and 4 of Chapter 4]. In general, most studies are based on the assumption that the surface potentials of the particle and the substrate remain constant as the distance of separation decreases. This assumption is valid only if the surface charge is generated by the potential determining ions and if a complete adsorption equilibrium is maintained during the interactions between the surfaces and the bulk solution. Experimental studies by Frens, Engel and Overbeek (1967), however, indicate that the rate at which adsorption equilibrium is established is not always sufficient to maintain the surface potentials constant. Thus, in such cases, the constant-potential assumption may be neither appropriate nor adequate. In addition, intermediate situations in which neither the surface potential nor the surface charge density remains constant during interaction can arise. These occur when charges are generated by the dissociation of acidic or basic groups on the surfaces (e.g., certain biological surfaces fall within this category). [See Section A (Double Layer Structure) and Section D (Forces Relevant to Biological Systems) of Chapter 3 for additional details of these phenomena.] These differences in the modes of double layer interaction have a direct impact on the rates of deposition and resuspension of colloidal particles, as illustrated by Figure 5.2 [taken from Rajagopalan and Kim (1981)]. In addition to their effect on adsorption phenomena, these differences in double layer interactions also exert a direct impact on the stability of colloidal dispersions, e.g., as measured by stability ratios of dispersions (see Chapter 4).

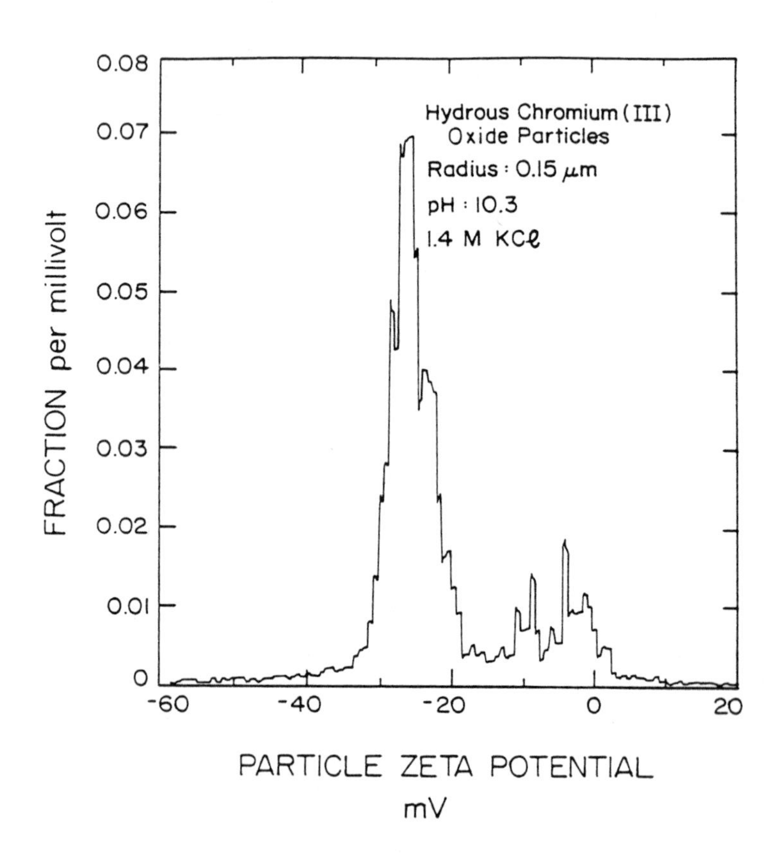

Figure 5.1 An example of the actual zeta potential distribution observed experimentally for a model colloid [from Rajagopalan and Chu (1982)].

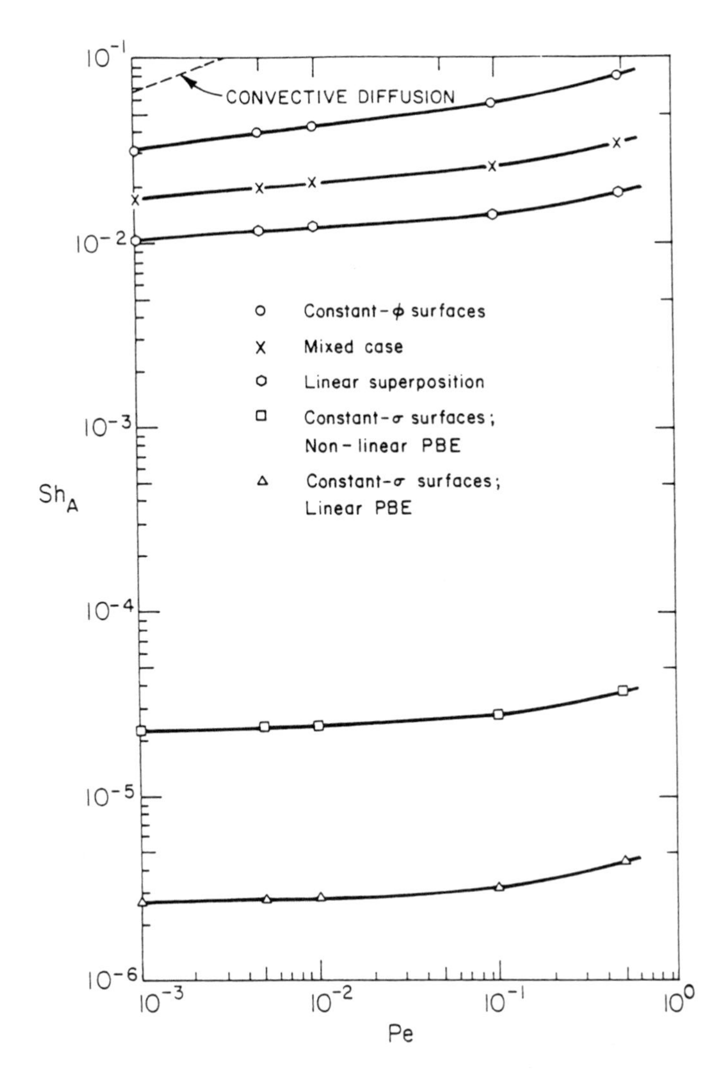

Figure 5.2 The deposition rate as a function of convection for different modes of double layer interaction [from Rajagopalan and Kim (1981)].

The deposition rate is represented by the Sherwood number, Sh_A, and convection is represented by the Peclet number, Pe. The indicated interaction modes are discussed in Chapters 3 and 4 (section A.3).

(vi) The rate expressions that are generally used to interpret experimental deposition and resuspension data are based on the assumption that convection near the surface (i.e., near the secondary minimum in the interaction potential) is not significant. This assumption, however, does not hold always, and the role of interfacial convection requires careful study [see Rajagopalan and Kim (1981)].

(vii) Many theoretical studies of deposition of colloidal particles onto substrates are confined to the initial deposition rate (i.e., when the particles begin to deposit onto clean substrates). In most applications, however, this initial period is only a small fraction of the total time involved. Consequently, the rate of deposition onto partially-loaded substrates and the mechanisms which affect this rate are very important. For example, as discussed later in this Chapter, experimental studies based on reasonably well-characterized substrates of simple geometry have shown that the adsorption rate declines very rapidly as particles begin to accumulate on the substrate, and, in general, much less than monolayer coverage is present when the substrate is entirely saturated.

(viii) As the deposition process continues beyond the initial period, the effects of particles already deposited on further deposition and on the rate must be taken into account. In addition, the extent and underlying causes (e.g., surface chemical changes and/or other factors) of the effects of already deposited particles need study so that these effects can be adequately quantified.

(ix) As will be described later, many of the theoretical and experimental studies of deposition have been based on model substrates of simple geometries. One important question, thus, is concerned with relating the experimental data and observations based on deposition on model, isolated substrates to deposition in complex geometries such as packed beds [see, for example, Rajagopalan and Tien (1976); Rajagopalan and Chu (1982)].

The above considerations make a purely theoretical or a purely experimental approach to the study of deposition and resuspension phenomena inadequate. Further, theoretical studies have attempted to focus on those features of the physical and chemical phenomena that can be extracted from experimental data despite the experimental uncertainties concerning the interaction potentials. For instance, the uncertainties regarding the exact mode of double layer interactions have led to theoretical studies on the effects of the general features of the interaction energy profiles (such as the shape of the curve and the magnitudes of the maximum and minima) and the role of interfacial convection. On the experimental side, better methods to produce model colloids of controllable physical and surface chemical characteristics are being investigated. In addition, experimental procedures which can monitor deposition and resuspension rates on model substrates in a non-intrusive manner are being developed. This Chapter does not aim to provide a comprehensive review of the literature on adsorption or deposition of colloidal species of all kinds on various substrates. Instead, we shall restrict the scope to a review of those theoretical and experimental studies that explore the role of colloidal forces on deposition and resuspension. Consequently, only those experimental studies that employ model colloids will be considered. In addition, although a part of this Chapter is devoted to deposition (and resuspension) in packed beds (because practical applications often involve deposition in porous media), our focus is again on model experiments and on the theories needed to interpret them. This excludes, for example, the vast amount of empirical information available in the literature on filtration in porous media. Experiments on deposition in packed beds offer certain advantages over those on model substrates such as rotating disks, and, hence, have been utilized extensively by Matijević and co-workers to study colloidal effects in filtration, corrosion, etc. (see Section B). Studies of this type are the focus of the discussion on deposition in complex geometries. However, information on specific problems, such as adsorption of microorganisms on soils, adsorption phenomena in biomedical and biochemical processes and the like, can be obtained from the books, monographs and reviews cited in the opening paragraph of this Chapter.

This Chapter is organized as follows. The deposition of colloidal particles onto model substrates (e.g., rotating disk, etc.) is discussed in Section A; both theoretical (Section A.1) and experimental results (Section A.2) are reviewed. The discussion for more complex geometries (e.g., packed beds) follows in Section B; again, both theory (Section B.1) and experiments (Section B.2) are presented.

Before we discuss these, a few general comments concerning the theoretical approach to the investigations of these phenomena are in order. The analysis of deposition or adsorption of colloidal species onto a substrate can be accomplished by dividing the process of deposition into two steps:

(1) a transport step, in which the particles are transported from the bulk of the solution to the substrate, and

(2) an attachment or adsorption step, in which they make contact with the substrate by overcoming any repulsive surface forces that may be important at short distances of separation.

The primary forces (and torques) which are relevant to these two steps may be classified into three categories:

(a) Forces and torques related to the motion of the fluid and the motion of the particles relative to the fluid and forces causing the Brownian motion of the particles; the hydrodynamic drag forces and torques and the 'diffusion forces' fall under this category;

(b) External forces such as those due to electric, magnetic and gravitation fields; and

(c) Chemical and colloidal forces which result from the interactions of substrates, particles and molecules and ions in the suspending medium; these include the van der Waals forces, coulombic forces and other surface forces mentioned previously.

The transport step is usually controlled by those in the first two categories. These forces are largely influential in bringing the particles to the substrate. Once the particles are close to the substrate, the forces in the third category begin to control the motion of the particles, especially when a strong repulsive barrier or a strong attraction exists. The attachment step is consequently dominated by these forces. Under certain conditions, however, convection in this region _can_ be important (see Section A.1 below).

This above outline provides an introduction to the discussions presented in the following sections, particularly in Sections A.1 and B.1, which focus on theoretical studies.

A. Deposition of Colloidal Particles on Model Substrates

Model substrates or collectors such as rotating disks or others such as spherical or cylindrical collectors or channels of different geometries (e.g., cylindrical channels) are used in both theoretical and experimental analyses of the mechanisms and rates of the deposition of colloidal particles on solid surfaces since the hydrodynamic flow fields associated with these model geometries are easy to characterize. Of these, the rotating disk collector has been the most widely used technique for studying the deposition of colloidal particles [e.g., Marshall and Kitchener (1966); Hull and Kitchener (1969); Clint et al. (1973); Dabros (1977); Adamczyk (1978); Tewari and Campbell (1978); Karis (1979); and Prieve and Lin (1980)]. The rotating disk technique now used in studies on colloidal interactions is an extension of a method originally developed for the study of the kinetics of electrochemical reactions [see, for example, Riddiford (1966) or Pleskov and Filinovskii (1976)]. In addition to the convenience of a relatively well-defined flow field, the rotating disk technique also has the advantage of generating a diffusion boundary layer of constant thickness. [The advantages of the rotating disk as an experimental technique are discussed in detail by Riddiford (1966); most of these also apply to studies of deposition of colloidal species.] Moreover, theoretical results obtained for deposition on rotating disks can be readily extended to other model substrate geometries such as spherical or cylindrical collectors, or to the case of heterocoagulation of particles of two different radii.

This Section is divided into two main parts; in the first (Section A.1), the theoretical approaches to the analysis of deposition of colloidal species on model substrates such as the rotating disk are discussed; and, in the second part (Section A.2), the experimental studies based on model collectors are briefly presented.

A.1 Theoretical Approaches to Adsorption of Colloidal Particles on Model Substrates

Levich (1962) originally derived an expression for the rate of mass transfer of ionic species to a rotating disk electrode when the reaction at the electrode is much faster than the diffusion of the species from the bulk of the liquid to the electrode; in the case of deposition of colloidal particles, this corresponds to situations in which the potential barrier against deposition is negligible. Several early studies on colloidal interactions used Levich's solution for the analysis of the observed rates of deposition [see Marshall and Kitchener (1966) and Hull and Kitchener (1969)]. Although some discrepancies between the Levich solution and the experimental results have been noted, it is generally agreed that this solution is sufficient for cases where the potential barrier is small (i.e., less than about 1 k_BT, where k_B is the Boltzmann constant and T the absolute temperature). On the other hand, for large potential barriers, drastic reductions in deposition rates have been observed, and appropriate modifications of the transport equation and its solution are needed to account for these decreased rates. One such modification is the approximate 'surface reaction approach' which is discussed below. In the general case, in which arbitrary interaction energy profiles prevail, this approximate method is insufficient and numerical solution of the appropriate transport equation is necessary.

As noted in the Overview to this Section, the process of deposition can be conveniently divided into two sequential steps, a transport step followed by an adsorption, or attachment, step. This classification, in fact, is useful in defining clearly the specific role played by surface and interfacial phenomena in the overall deposition process. Trajectory analyses of the deposition process have shown that the role of surface interactions on the transport of the particle from the bulk of the suspension into the vicinity of the substrate or collector is quite small [see Rajagopalan and Tien (1977, 1979)]. However, as soon as the particle is close to the substrate, the role of the surface interactions is paramount and these govern the attachment step.

The surface forces generally operate over very short distances from the substrate (approximately 100 nm). Thus, there exist, in effect, two regions (an outer region and an inner region) characterized by the relative magnitude of forces (see Figure 5.A.1). The convective and the other transport components of the diffusion equation, which exert a strong influence in the outer region, may be conveniently omitted in the inner region (i.e., for distances from the substrate less than about 100 nm) for cases in which the interaction potential has a large barrier and a low secondary minimum; this inner region is termed the 'surface interaction boundary layer'. Several investigators have used this division, in what may be called a *surface reaction model*, to describe the deposition of colloidal particles [see, for example, Ruckenstein and Prieve (1973); Wnek (1973); Spielman and Friedlander (1974); Dahneke (1974); Dahneke (1975a,b); and Zimmer and Dahneke (1976)]. This surface reaction model can be briefly described as follows. The solution of the convective diffusion equation for the outer region (without the surface interaction components) is matched appropriately with the solution of the diffusion equation (with surface interaction but without convection) in the inner region, i.e., inside the surface interaction boundary layer. As shown by Ruckenstein and Prieve (1973), the net flux to the surface under these conditions using this matching procedure can be mathematically represented by

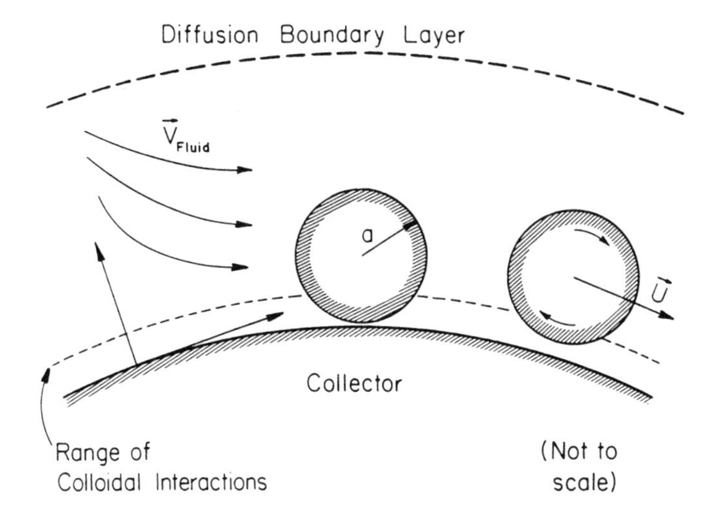

Figure 5.A.1 Schematic representation of the region close to a collector in terms of forces important in the various segments of the region.

a fictitious first-order surface reaction, the rate coefficient of which is a function of the interaction potentials. Thus, the surface of the substrate appears as a sink where a first-order chemical reaction constantly consumes the arriving particles; and the convective diffusion equation for the outer region (without the surface interaction components) can be solved in combination with this surface reaction boundary condition to obtain the deposition rate. For additional details, refer to the references cited previously; see also Ruckenstein and Prieve (1980); in addition, this theoretical approach is summarized in Rajagopalan and Tien (1979) and in Rajagopalan and Karis (1979). Solutions based on this surface reaction model have also been presented by Bowen, Levine and Epstein (1976) for describing particle deposition in parallel-plate and cylindrical channel flows. Additional results (both theoretical and experimental) are reported in Bowen and Epstein (1979) for deposition in parallel-plate channels. Details of their experimental results will be presented later, in Section A.2.

The advantage of the above approach is that the solution can be obtained in an analytical (but, approximate) form for simple geometries (e.g., rotating disks). However, it should be emphasized that the surface reaction model is valid only when the interaction potential displays a large barrier (greater than about 5 k_BT) and a low secondary minimum (less than about 0.5 k_BT), as described below. In addition, the surface interaction boundary layer must be very much thinner than the diffusion boundary layer [see, for example, Prieve and Ruckenstein (1976) and Karis (1979)]. This has been known, at least qualitatively, for some time. More recently, Dabros and Adamczyk (1979) have provided some quantitative criteria, based on computations of rates of adsorption on rotating disks. Their calculations indicate that the surface reaction model is sufficiently accurate for situations in which the Peclet number is less than approximately 10^{-2} and in which the potential barrier is greater than about 5 k_BT. (Note that an increase in the magnitude of the Peclet number implies that the influence of convection in the surface interaction boundary layer increases.) These criteria, moreover, are strictly valid only when the depth of the secondary minimum in the interaction potential is rather low, i.e., less than approximately 0.5 k_BT, as stated earlier.

Furthermore, the calculations of Dabros and Adamczyk are based on a particular choice of the interaction potential expression that is valid when the interacting surfaces maintain constant surface potentials as they approach each other. (In fact, the actual expression used by Dabros and Adamczyk is a simplified form of the appropriate interaction potential.) However, as was discussed in Chapters 3 and 4, situations in which other modes of interactions (e.g., constant charge density interactions) are physically more reasonable do arise frequently in practice. A change in the mode of interaction can lead to significant changes in the magnitudes of the interaction energy barrier and the depth of the secondary minimum. These, in turn, can cause large inaccuracies in the deposition rates calculated on the basis of the surface reaction model for reasons discussed in the previous paragraphs. To avoid such inaccuracies, one is often forced to seek numerical solutions to the relevant transport equations. For these same reasons, most of the theoretical studies in the literature on particle adsorption or deposition are based either on assuming simple expressions for the intermolecular forces [e.g., Dabros and Adamczyk (1979)] or on a simultaneous solution for the forces of interaction and their influence on adsorption [e.g., Wnek, Gidaspow and Wasan (1977)].

To circumvent the need to calculate deposition rates separately for each mode of interaction, Rajagopalan and Kim (1981) have recently developed closed-form equations, based on approximate asymptotic solutions and exact numerical results, which can be used regardless of the particular mode of interaction as long as the assumptions needed for the surface reaction model hold. When these assumptions are not applicable, interfacial convection becomes significant, as demonstrated quantitatively in the above paper. Rajagopalan and Kim (1981) demonstrate that the surface chemical phenomena and the purely physical aspects of the adsorption process can, in fact, be separated conveniently so that the adsorption rate can be related to general classes of surface interaction potentials. These classes of surface interaction potentials, in turn, can be specified sufficiently accurately in terms of two or three parameters (e.g., height of the barrier, Hamaker constant, etc.). Thus, there is no need to associate the adsorption rate with individual physicochemical properties and the mode of double layer interaction. These results can be briefly reviewed as follows, with reference to the net interaction energy profiles shown in Figure 5.A.2 [taken from Kim and Rajagopalan (1982)]. These potential profiles represent the sum of coulombic and van der Waals interaction energies between two surfaces. Curve I (labeled 'Type I') exhibits a large positive primary maximum (i.e., V_{max}) and a negligible secondary minimum ($V_{s,min}$). The deposition rate in this case is determined primarily by the height of the potential barrier, the magnitude of the attractive London-van der Waals force (represented by the London group, N_{Lo}), and the strength of convection (represented by the Peclet number). Curve II (labeled 'Type II') displays, in addition to a large positive V_{max}, a substantial secondary minimum. The presence of this secondary minimum gives rise to an accumulation of particles near the secondary minimum; this accumulation tends to increase

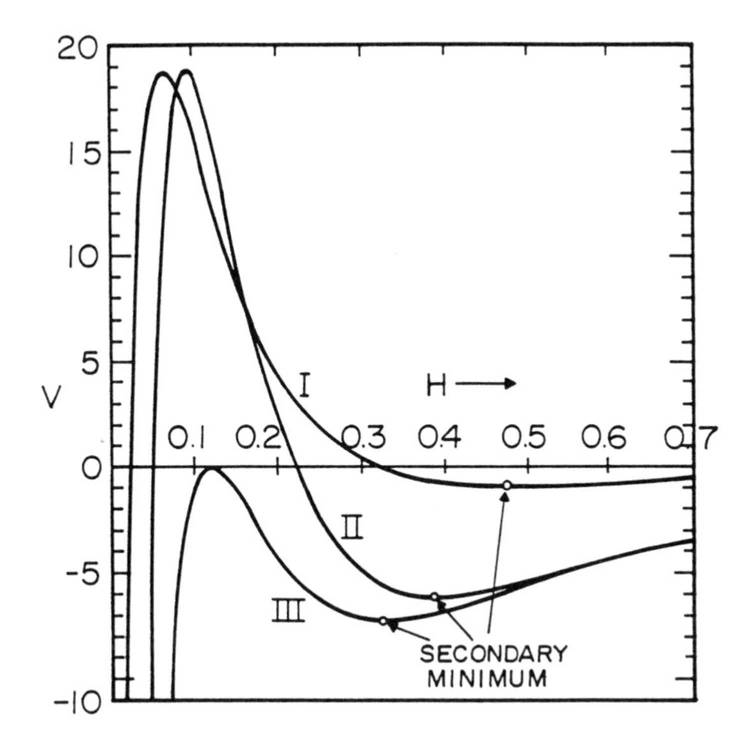

Figure 5.A.2 Three classes of shapes of the interaction energy profiles [from Kim and Rajagopalan (1982)].

The deposition rate and its dependence on interfacial convection and energy barrier are different for these different classes of profiles.

the convective transfer of particles parallel to the substrate if there is a significant velocity component in this direction. Thus, although the deposition rate in the case of Type II potentials is still strongly influenced by the large potential barrier, the rate is also sensitive to the depth of the secondary minimum. As a result, in the case of Type II potentials, the deposition rate is a function of four parameters, i.e., V_{max}, $V_{s,min}$, N_{Lo} and the Peclet number. The third profile, Curve III (and labeled 'Type III') corresponds to profiles with a numerically low value of V_{max}, and a deep secondary minimum. In this case, there is competition between the normal flux and the tangential flux since the resistance due to the potential barrier is no longer dominant; the flux, in fact, first increases and then decreases as the Peclet number increases [Rajagopalan and Kim (1981); Kim and Rajagopalan (1982)]. In addition to these, it is also convenient to consider a fourth class of interaction energy profiles resulting from purely attractive forces. Recently, closed form analytical solutions for this class of interaction profiles have been obtained by Chari (1984) [see, also, Chari and Rajagopalan (1984)].

It should be emphasized that the _chemical_ nature of the double layer interaction is already implicitly accounted for in the analysis described above since the major effect of this is the resultant change in the _shape_ of the profile, and since the _physics_ of the adsorption process is completely determined by the shape. A detailed discussion of these, related results and numerical examples are presented in Rajagopalan and Kim (1981). That paper also presents a closed-form equation for the rate of adsorption, as a function of all relevant parameters, for Type I potentials. The model problem in that study was deposition of colloidal particles on rotating disks. Another study

[Kim and Rajagopalan (1982)] applies the concepts described above to more practical cases; e.g., adsorption on cylindrical and spherical collectors and in packed beds and mutual coagulation.

Other significant conclusions regarding the deposition process drawn from the studies above include the following.

(i) In the absence of repulsion, the adsorption rate is determined by the magnitudes of the London group and the Peclet number. The attractive van der Waals force exerts generally very little influence on the transport of the particle to the surface of the collector, as discussed previously. However, as is well-known, this force is needed in the calculations to offset the hydrodynamic retardation of the particle motion.

(ii) For situations in which interaction energy profiles of Type II or Type III (see Figure 5.A.2) are dominant, three factors influence deposition rate (or stability ratio); these are the 'activation energy' [i.e., this activation energy is the difference between the values of potential at the primary maximum and at the secondary minimum (see Figure 5.A.3)], the extent of convection, and the depth of the secondary minimum. In addition, when the primary minimum is not too deep, reversible adsorption and coagulation can occur [e.g., Rajagopalan and Kim (1981)]. The effect of the shape of the interaction energy profiles on these above phenomena has also been studied by Ruckenstein (1978); Ruckenstein has established conditions under which the rates of reversible coagulation in the primary minimum can be computed by neglecting accumulation in the secondary minimum.

While the papers cited above focus on general classes of interaction profiles and their influence on deposition, other investigations, with a focus on specific interaction profiles and on selected model substrates, have also been reported in the literature, most notably by Adamczyk, Dabros, van de Ven and co-workers. These investigators have studied, both theoretically and experimentally, the deposition of particles onto rotating disks [for example, Dabros (1977); Dabros, Adamczyk and Czarnecki (1977); Adamczyk (1978); Dabros and Adamczyk (1979); and Adamczyk and van de Ven (1983)] and other collectors such as cylindrical collectors [e.g., Adamczyk and van de Ven (1981b)] or in parallel-plate and cylindrical channels [e.g., Adamczyk and van de Ven

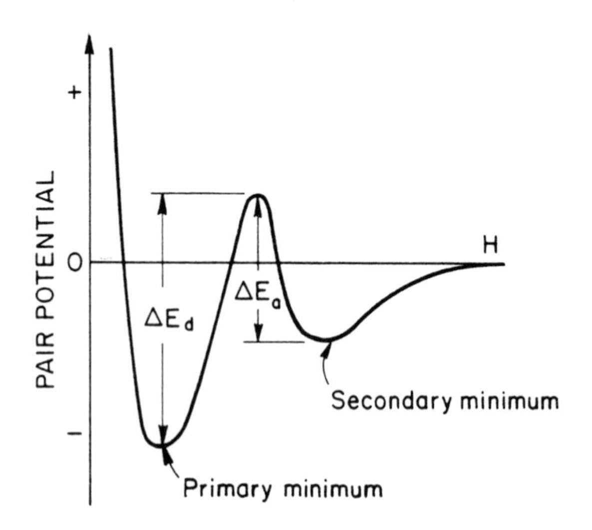

Figure 5.A.3 The activation evergy for adsorption, ΔE_a, and for desorption, ΔE_d.

When ΔE_d is large, desorption (i.e., resuspension) can be assumed to be zero. For low values of ΔE_d, reversible adsorption can occur.

(1981a)]. In these and related theoretical studies, the transport equations corresponding to the given situation and collector geometry are formulated; these equations take into account specific surface interactions (i.e., dispersion and double layer forces) as well as external forces (e.g., gravity). A recent paper by Adamczyk, Dabros and van de Ven (1983) reviews the formulation of the transport equations for selected model geometries and discusses problems related to the choice of appropriate boundary conditions for solution of the transport equations. This review also presents a general theory to describe the deposition of colloidal particles in terms of a 'mobile' phase and an 'immobile' phase (in which particles are immobilized at the surface of the substrate). In this model, the perfect sink boundary condition arises as a consequence of being the limiting case of a fast immobilization reaction. For additional details, see the review by Adamczyk, Dabros and van de Ven (1983), and references cited therein. (Note also that deposition of colloidal particles on a rotating disk collector is discussed in detail therein.) A review of the status of theory for interactions between two particles in shear flow, both in the absence and in the presence of colloidal forces, is given in van de Ven (1982).

In addition, van de Ven and co-workers have recently reported results based on a direct experimental technique for studying particle deposition on transparent solid surfaces under conditions of a stagnation point flow [e.g., Dabros and van de Ven (1982)]. Theoretical results for predicting the corresponding mass transfer rates in stagnation point flow, as well as results of deposition experiments, are described in this paper. Other theoretical studies are also in progress [see, for example, Dabros and van de Ven (1983a,b); Adamczyk and van de Ven (1983)]. For additional details and references, see the review by Adamczyk, Dabros and ven de Ven (1983).

Before concluding this discussion of theoretical analyses of deposition on model collectors, it is worth noting that some additional results, for deposition on collectors other than rotating disks, are available in a few papers cited earlier and in a few others cited below. These include theoretical studies for spherical collectors [e.g., Prieve and Ruckenstein (1974); Kim and Rajagopalan (1982)], and cylindrical collectors [e.g., Adamczyk and van de Ven (1981b); Kim and Rajagopalan (1982)] and for deposition onto walls of parallel-plate channels [e.g., Bowen, Levine and Epstein (1976); Bowen and Epstein (1979); Adamczyk and van de Ven (1981a)], or cylindrical channels [e.g., Bowen, Levine and Epstein (1976); Adamczyk and van de Ven (1981a)].

A.2 Experimental Studies of Deposition of Colloidal Species on Model Substrates

Model colloids have been used in experimental studies of colloidal stability [see Chapter 4] as well as in studies of the deposition of colloidal particles onto model, or other, substrates. Methods for producing model colloids of various sizes and for controlling or adjusting their surface potentials or other properties have become increasingly available in recent years and are now well-established [see, for example, Demchak and Matijević (1969); Matijević and Scheiner (1978); Visca and Matijević (1979); Sugimoto and Matijević (1980); Sapieszko and Matijević (1980); Matijević (1981); Milić and Matijević (1982); Matijević and Wilhelmy (1982); Bagchi, Gray and Birnboum (1979); and Partch et al. (1983)].

The majority of the reports on deposition experiments available in the literature, as well as most of the theoretical studies as described in Section A.1 above, are confined to the <u>initial</u> deposition rate (i.e., when the particles begin to deposit on clean substrates). For such initial deposition phenomena, as noted above, the perfect sink boundary condition applies and particle removal, or desorption, from the substrate can be neglected. In this case, experimental results are, in general, in good agreement with the theoretical predictions of deposition rates when surface conditions are favorable. Experiments have been performed for deposition of colloidal particles onto model collectors such as rotating disks [see the references cited on rotating disk studies in Section A.1; also Prieve and Carrera (1978); Prieve and Lin (1980)], cylindrical collectors [e.g., Adamczyk and van de Ven (1981b)], and channels of parallel-plate or cylindrical or other geometries [e.g., Hermansson (1977); Bowen and Epstein (1979); Adamczyk and van de Ven (1981a)]. Other experimental approaches to this class of problems also include studies of adsorption of small dispersed particles on larger particles [e.g., Bleier and Matijević (1978); and Hansen and Matijević (1980)]. Recently, as discussed in Section A.1 above, van de Ven and co-workers have developed a stagnation flow point system for studying the deposition process, both theoretically and experimentally. The experimental procedure and results of deposition experiments are described in Dabros and van de Ven (1982); see also van de Ven, Dabros and Czarnecki (1983) and references therein.

Some of the major observations and results of the available experimental studies of the deposition process are highlighted below.

(i) As noted in the Overview to this Chapter, deposition experiments have demonstrated that the rates of adsorption and desorption are sensitive to the chemical types of ions in the suspension, the zeta potentials, ionic association/dissociation equilibria, the presence of impurities in the suspensions (even at very low concentrations), the mode of interaction of the double layers (e.g., constant surface charge density, constant surface potential, or a combination of these), among other physicochemical variables.

(ii) Experimental evidence has shown that the rates of deposition agree well with those predicted by the theory, when the particles and the substrate are <u>oppositely</u> charged.

(iii) On the other hand, when the particles and substrates are of the same charge, the observed deposition is much slower and the observed rate of deposition is much less than that which would be predicted by theory. [This effect may, in part, be due to the effects of surface roughness and surface charge or chemical heterogeneities on the magnitude of the energy barrier. Additional studies are needed to understand these effects.]

(iv) As was discussed previously, experimental studies of the deposition process have demonstrated that the rate of deposition falls off rapidly and non-linearly with time. Particles which have been previously deposited on the substrate alter the geometry of the collector surface, as well as the flow in the region near the collector. In general, as particles accumulate, these effects lead to a decrease in the deposition rate; and there is usually much less than a monolayer coverage when the substrate is saturated [e.g., see Rajagopalan and Chu (1982); Adamczyk, Dabros and van de Ven (1983)]. Other investigators have reported that in some cases the variations in the rate of deposition with quantity of adsorbed species are due to a change in the charge of the substrate [e.g., Wnek, Gidaspow and Wasan (1975, 1977); and Gregory and Wishart (1980)].

(v) In addition to this change in the surface chemical state or charge of the substrate as the deposition process continues in time [i.e., see (iv) above], another possible cause of the observed eventual decrease in the adsorption rate is the desorption of already adsorbed particles. Studies of the desorption of colloidal particles from selected substrates and/or the conditions under which desorption occurs are reported in, for example, Bowen and Epstein (1979); Kolakowski and Matijević (1979); Matijević (1980); Kuo and Matijević (1979, 1980); Kallay and Matijević (1981); and Kallay, Nelligan and Matijević (1983).

(vi) The observed decreases in the rate of deposition and in maximum surface coverage as deposition continues can be explained, in part, by what can be termed an 'excluded-area' effect which arises due to the finite size of the particles that are already adsorbed. That is, a particle already adsorbed on the substrate reduces the area that is available for further adsorption by an amount that is generally larger than its projected area [e.g., see Rajagopalan and Chu (1982)]. This effect has been termed the 'shadow effect' by Tien, Wang and Barot (1977) in their studies on aerosol deposition. In the case of colloids, this effect is enhanced by the presence of the ionic cloud which surrounds the adsorbed particle. Additional details and techniques to estimate this shielded area are described in Rajagopalan and Chu (1982). This 'shadow effect' has also been studied by van de Ven and co-workers, who call this phenomenon 'blocking' (by already deposited particles). For details on their formulation of this problem, see Adamczyk, Dabros and van de Ven (1983).

(vii) The decrease in deposition rates can also be caused by coagulation of the particles in the bulk, although this has not been investigated so far.

The experimental studies reviewed above were concerned primarily with the deposition of colloidal species onto selected _model substrates_. However, many of these conclusions apply also to deposition of colloidal particles onto substrates of more complex geometries. A special case of the latter, which has received considerable attention in the literature because of its value in practical applications, is deposition in porous media such as packed beds; this is discussed next in Section B.

B. Adsorption of Colloidal Particles in Packed Beds

The theoretical and experimental results described in the preceding Section A are based on studies of adsorption of colloidal species onto model substrates. The advantages of such collectors include their simple geometries and the fact that the hydrodynamics associated with these are well-characterized. However, it is not straightforward to relate such studies of model collectors to substrates of more complex geometries and features. Thus, one possible advantage of studying deposition phenomena in a complex system such as a packed bed is that the results obtained are applicable to many situations of practical interest more directly. (Such an example is filtration in packed beds.) The use of packed beds and porous media for the physical separation of colloidal particles dispersed in liquid media has been known for some time. Originally, this method of separation was thought to be governed by only mechanical mechanisms of separation; and, consequently, the mechanics of the deposition process was the first aspect of the phenomenon to attract the interests of researchers in this area. [An extensive review of the modern theories of packed bed filtration is available in Rajagopalan and Tien (1979).]

However, as was discussed in Section A, the deposition of colloidal species onto substrates is critically influenced by the physicochemical interactions between the suspended particles and the substrates. In the case of complex collectors such as packed beds, other effects may also be important. For instance, the complex geometry of the pores generally promotes other physical mechanisms of deposition such as interception or sedimentation [see, for example, Yao, Habibian and O'Melia (1971)]. Consequently, one might expect that a combination of the various physical mechanisms and the surface chemical effects to control the rate. Other questions of relevance to studies of deposition in complex collectors include the following issues.

(i) The extension of observations based on experiments with model substrates of well-defined geometry to more complex situations that prevail in the pores of packed beds is not well-understood.

(ii) The extent to which particles already deposited in the pores of a packed bed affect subsequent deposition and the rate is one subject of current studies [see also the discussion in Section A of this Chapter on the effects of already adsorbed particles on subsequent deposition].

(iii) Procedures for predicting *a priori* the breakthrough (see Figure 5.B.1) and related behavior of a packed bed operation require further study. Available procedures for estimating the capacity of the substrate, for example, need additional investigation since the excluded-area effect would be expected to depend on physicochemical properties such as pH, ionic strength and others [e.g., Rajagopalan and Chu (1982)].

Other questions are considered in the discussion below. As in Section A, this section is divided into two parts; the first presents a review of theoretical studies of deposition of colloids in packed beds, and the second presents a brief review of relevant experimental studies.

B.1 Theoretical Analysis of Adsorption in Packed Beds

Adsorption and desorption of colloidal particles in packed beds can be treated theoretically either from a 'macroscopic' or from a 'microscopic' point of view. The macroscopic, or phenomenological, approach is based on a mass balance on an elemental slice of the bed and follows the overall, average concentration history at any arbitrary location in the bed. In contrast, the microscopic approach attempts a closer scrutiny of the individual processes that occur in the interstices of the bed, through a sequence of theoretical procedures, each of which focuses on one aspect of the problem [see Rajagopalan and Tien (1976, 1979, 1982)]. The advantage of this approach is that the rates of adsorption and desorption are accounted for in terms of the

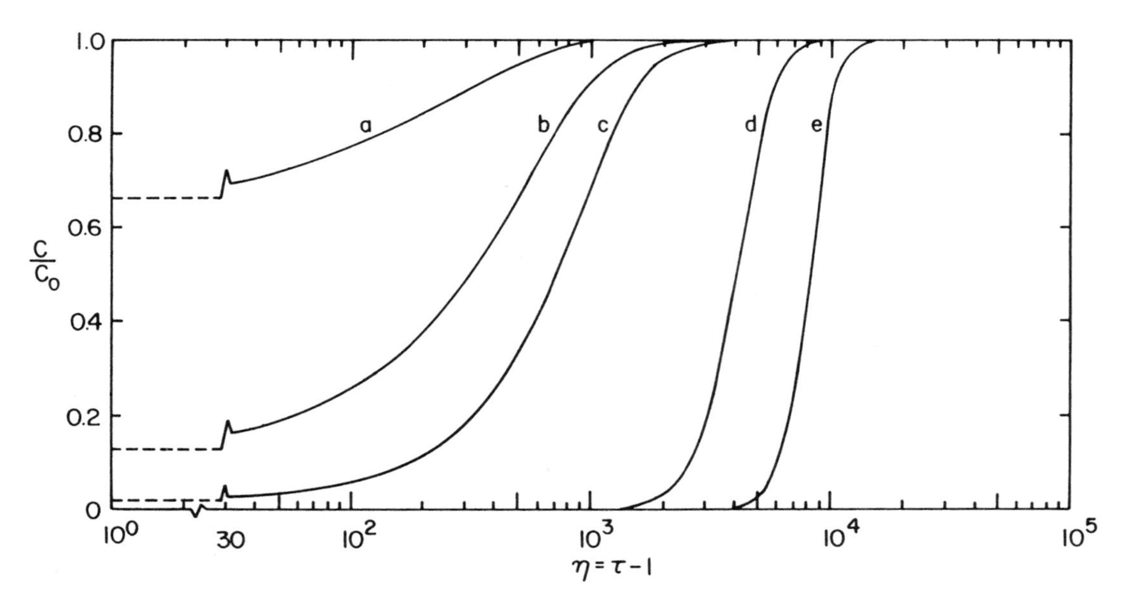

Figure 5.B.1 Concentration of particles in the suspension at the exit of a packed bed [from Rajagopalan and Chu (1982)].

The concentration C_o is the concentration at the inlet; τ is the dimensionless time; η is the dimensionless 'elasped' time at the exit. The curves shown are for increasing rates of deposition in the bed (the rate is the lowest for 'a' and the highest for 'e').

relevant mechanisms (such as transport, adsorption and interception), whereas in the first (i.e., macroscopic) approach, these rates are accounted for in terms of undefined, or empirical, mass transfer coefficients. However, since one usually measures only the concentration history in packed bed experiments, it is essential that the microscopic approach be integrated with the macroscopic approach in order to compare their predictions with experimental results. These two methods are discussed in detail in Rajagopalan and Tien (1979).

A theoretical formulation which combines these two approaches above for the problem of interpreting experimental breakthrough curves for adsorption of colloidal particles in a packed bed is presented in Rajagopalan and Chu (1982). This model can be described briefly as follows. A mass balance on the flowing suspension in an elemental slice of the packed bed results in the usual continuity equation; axial and radial dispersions are assumed to be negligible, as is the radial variation of concentration. The local flux as a function of time is represented by a simple rate equation in which two rate coefficients represent adsorption (forward rate coefficient) and desorption (reverse rate coefficient). These rate coefficients must be specified as functions of the parameters which influence the deposition and resuspension mechanisms. Methods to obtain expressions for the rate coefficients for adsorption and desorption are described in this paper; also, see Dahneke (1975) for discussion on expressions for the desorption rate coefficient. The net local flux has been assumed by Rajagopalan and Chu (1982) to be the difference between the forward rate and the reverse rate. The forward rate is taken to be linearly proportional to the bulk concentration; similarly, the reverse rate is assumed to be linearly dependent on the particle concentration at the surface. Note that this form is common in the theory of linear chromatography [e.g., see Aris and Amundson (1973)] and, hence, is referred to as the 'linear chromatography approximation' by Rajagopalan and Chu (1982). The advantage of this procedure is that closed-form relations for the breakthrough curves, surface concentrations and local rates (inside the bed at any position and at any time) can be obtained in terms of the rate coefficients, which are determined by surface chemical conditions. Computed results are presented in Rajagopalan and Chu (1982), and a typical set of breakthrough curves is shown in Figure 5.B.1. Rajagopalan and Chu also relate the eventual decrease in adsorption rate (as the grains of the bed become saturated) to the excluded area effect described previously in Section A of this Chapter. The procedure presented in this paper is adequate for estimating time of breakthrough, length of bed and other design parameters for packed bed filters. Other studies are needed, however, to investigate the sensitivity of the excluded area effect to physicochemical properties such as pH, and so forth, as noted earlier. For additional discussion of this formalism, see the paper by Rajagopalan and Chu (1982).

It has been suggested previously [see Wnek, Gidaspow and Wasan (1975)] that slow accumulation of particles in the pores gradually shifts the interaction to a less favorable one. The assumption here is that the observed reduction in flux can be traced to the ionic (charge) concentrations on the grains of the packing because of the continual arrival of particles with charges of the opposite sign. This change in charge concentration has been incorporated in the transport equations by Wnek, Gidaspow and Wasan (1975). The computations based on this approach, however, are not consistent with the corresponding local rates, although the breakthrough time can be estimated with sufficient accuracy. In addition, the charge balance approach leads to equations that cannot be solved analytically, whereas the linear chromatography approximation permits analytical solution. The charge balance concept can be incorporated, semi-empirically, in the linear chromatography approach by relating the reverse rate constant to the surface saturation concentration, which is determined on the basis of charge neutralization of the substrate; see Rajagopalan and Chu (1982).

B.2 Experimental Studies of Adsorption of Colloidal Particles in Packed Beds

Experimental studies of deposition of colloidal species in packed beds have been concerned with investigations of the effects of various parameters such as surface charge, ionic strength, pretreatment procedures, particle size, pH, etc. on the adsorption or desorption of particles from surfaces such as steel or glass. These experimental studies include, for example, those of Heertjes and Lerk (1967); Wnek, Gidaspow and Wasan (1975); and more recently, Rajagopalan and Chu (1982). [Additional studies are referenced in these papers.] Matijević and co-workers have also reported many experimental studies of deposition of colloidal particles in packed beds [e.g., Kolakowski and Matijević (1979); Kuo and Matijević (1979, 1980); Kallay and Matijević (1981); Matijević, Kuo and Kolny (1981); Nelligan, Kallay and Matijević (1982); Kallay, Nelligan and Matijević (1983); Thompson, Kallay and Matijević (1983); Matijevic and Kallay (1983); and references therein].

In contrast to the dynamic packed bed studies of Heertjes and Lerk (1967), Wnek, Gidaspow and Wasan (1975) and Rajagopalan and Chu (1982), the experiments of Matijević and co-workers do not attempt to measure the breakthrough curves. Instead, these investigators take advantage of the large surface area afforded by the packing material to measure deposition and resuspension rates over specified periods of time. These rates are, then, compared with predictions based on surface reaction models.

The major results and conclusions of these packed bed studies include the following observations (see also Section A.2 above).

(i) The rate of deposition declines rapidly as surface coverage increases even in instances in which the surface forces were favorable for deposition. This decline is due largely to an excluded area effect introduced by the already adsorbed particles [e.g., Rajagopalan and Chu (1982); and Kallay, Nelligan and Matijević (1983)]. The excluded area effect was discussed above, in Section A.2.

(ii) Based on analysis of experimental data, it appeared that this decrease in the local deposition rate could be expressed adequately in terms of a desorption term in the rate equation.

(iii) The adsorbed particles do not readily desorb unless the physicochemical properties of the liquid are altered in such a manner that a net repulsion develops between the particle and substrate. Kuo and Matijević (1980) reported results which indicate the strong dependence of desorption on pH and ionic strength of the liquid [see also Kallay, Nelligan and Matijević (1983)].

Other results and details may be obtained from the papers cited above.

Before concluding this section, a few remarks regarding deposition in the case of nonaqueous colloids are appropriate. The discussions presented in Sections B.1 and B.2 on the work of Rajagopalan and Chu were based on experiments and studies of deposition of aqueous colloids. However, the conclusions of Rajagopalan and Chu (1982) are consistent with the results obtained for nonaqueous colloids by Chowdiah, Wasan and Gidaspow (1981), who investigated filtration of carbon particles in tetralin through sand beds. Experiments based on model nonaqueous colloids, however, are not available in the literature.

Research Needs

The eventual solutions of problems related to mass transfer in colloidal systems depend, in part, on progress in resolving the uncertainties associated with the various electrokinetic and intermolecular forces as discussed in Chapter 3. These include, for example, research needs in advancing the understanding of electrical double layer structure and double layer interactions under dynamic conditions, expressions for attractive forces, and the short-range structural forces. In general, a better understanding and description of deposition and resuspension phenomena requires additional research into the nature, both qualitative and quantitative, of particle-particle or particle-surface interactions at very small distances of separations (see also the Research Needs listed at the end of Chapter 3). In addition, other research needs in the area of deposition on substrates include the following.

- Understanding the role of physical and chemical factors on the deposition and resuspension of particles is complicated by the interdependent nature of these variables (e.g., zeta potential, types of ions in solution, association/dissociation phenomena, etc.). Furthermore, some of these parameters are not simple to quantify sufficiently accurately. Additional theoretical and experimental investigations are needed to address these issues, as well as to study the sensitivity of deposition phenomena to these parameters.

- The role of interfacial convection in deposition and resuspension phenomena needs further study since the assumption that convection near the surface is insignificant is not always valid.

- Systematic studies of preparing clean surfaces or surfaces of uniform properties are needed in order to interpret experimental data reliably. These are especially needed in light of the fact that surface heterogeneities have been known to be important for a long time, and are difficult to control.

- Some investigators have suggested that the discrepancies between theoretical predictions of adsorption rates and experimental observations may be due to coagulation of particles in the bulk (see Section A of this Chapter). Adsorption experiments that simultaneously monitor coagulation in the bulk will be useful in resolving this issue.

- Another outstanding research problem is related to the analysis of deposition and deposition rates onto substrates for times beyond the initial deposition period. Additional theoretical and experimental studies of deposition onto partially-loaded substrates which can take into account the effects of already deposited particles on subsequent adsorption and desorption of particles are needed.

- Concentration buildups near the substrates could severely affect the deposition rate and are important in numerous practical systems such as cake filtration and membrane filtration. This phenomena requires further study; related problems of concentration effects are discussed in Chapter 6.

- As noted previously in this Chapter, results of studies on model substrates of well-defined geometries are not easily extrapolated to more complex geometries. Additional studies are thus needed to examine this issue and to determine, for example, methods to predict a priori the breakthrough behavior in a complex geometry such as a packed bed.

- As discussed in the Research Needs given at the end of Chapter 3, the application of principles of colloid science to biocolloids is just beginning. The research needs in problems related to biological adsorption are immense due partly to the fact that biocolloidal systems are widely encountered in both natural and industrial processes. Thus, there is a great need for theoretical and experimental studies on the nature and rates of the mechanisms relevant to adsorption and desorption of biological species to surfaces.

REFERENCES: OVERVIEW

Anfinsen, C. B., Edsall, J. T. and Richards, F. M. (eds.), Advances in Protein Chemistry, Academic Press, New York, NY, 1978.

Berkeley, R. C. W., Lynch, J. M., Melling, J., Rutter, P. and Vincent, B. (eds.), Microbial Adhesion to Surfaces, Wiley, New York, NY, 1981.

Bitton, G. and Marshall, K. C. (eds.), Adsorption of Microorganisms to Surfaces, Wiley-Interscience, New York, NY, 1980.

Frens, G., Engel, D. J. C. and Overbeek, J. Th. G., Trans. Faraday Soc. 63, 418 (1967).

Giaever, I. and Ward, E., Proc. Nat. Acad. Sci. 75, 1366 (1978).

Grinnell, F., Int. Rev. Cytology 53, 65 (1978).

Kavanaugh, M. C. and Leckie, J. O. (eds.), Particulates in Water, American Chemical Society, Washington, D. C., 1980.

Litwin, S. D., Meth. Immunology Immunochem. 4, 115 (1977).

Lyklema, J., Pure Appl. Chem. 52, 1221 (1980).

MacRitchie, F., pp. 283-326 in Anfinsen, Edsall and Richards (1978).

Marshall, K. C., Interfaces in Microbial Ecology, Harvard Univ. Press, Cambridge, Mass., 1976.

Matijević, E., Kuo, R. J. and Kolny, H., J. Colloid Interface Sci. 80, 94 (1981).

Murray, J. P. and Parks, G. A., pp. 97-133 in Kavanaugh and Leckie (1980).

Rajagopalan, R. and Chu, R. Q., J. Colloid Interface Sci. 86, 299 (1982).

Rajagopalan, R. and Kim, J. S., J. Colloid Interface Sci. 83, 428 (1981).

Rajagopalan, R. and Tien, C., AIChE J. 22, 523 (1976).

Rajagopalan, R. and Tien, C., Canad. J. Chem. Engr. 55, 246 (1977).

Rajagopalan, R. and Tien, C., pp. 179-269 in Wakeman (1979).

Rajagopalan, R., Tien, C., Pfeffer, R. and Tardos, G., AIChE J. 28, 871 (1982).

Sasaki, H., Matijević, E. and Barouch, E., J. Colloid Interface Sci. 76, 319 (1980).

Somerscales, E. F. C. and Knudsen, J. G. (eds.), Fouling of Heat Transfer Equipment, Hemisphere Publishing, Washington, D. C., 1981.

Wakeman, R. J. (ed.), Progress in Filtration and Separation, Vol. 1, Elsevier, Amsterdam, The Netherlands, 1979.

Wnek, W. J., Gidaspow, D. and Wasan, D. T., Chem. Eng. Sci. 30, 1035 (1975).

REFERENCES: SECTION A. DEPOSITION OF COLLOIDAL PARTICLES ON MODEL SUBSTRATES

Adamczyk, Z., Ph.D. Dissertation, Polish Academy of Science, Krakow, Poland, 1978.

Adamczyk, Z., Dabros, T. and van de Ven, T. G. M., pp. 619-635 in Poehlein, Ottewill and Goodwin (1983).

Adamczyk, Z. and van de Ven, T. G. M., J. Colloid Interface Sci. 80, 340 (1981a).

Adamczyk, Z. and van de Ven, T. G. M., J. Colloid Interface Sci. 84, 497 (1981b).

Adamczyk, Z. and van de Ven, T. G. M., Kinetics of Particle Accumulation at Collector Surfaces. I. Approximate Analytical Solutions, PGRL Report No. 268, Pulp and Paper Research Institute of Canada, Pointe Claire, Canada, 1983.

Bagchi, P., Gray, B. V. and Birnboum, S. M., J. Colloid Interface Sci. 69, 502 (1979).

Becher, P. and Yudenfreund, M. (eds.), Emulsions, Dispersions and Latices, Marcel Dekker, New York, NY, 1978.

Beddow, J. K. and Meloy, T. P. (eds.), Testing and Characterization of Powders and Fine Particles, Heyden, London, U. K., 1980.

Bennett, G. F. (ed.), Water--1978, AIChE Symposium Series Vol. 75, Amer. Inst. Chem. Engr., New York, NY, 1979.

Bleier, A. and Matijević, E., J. Chem. Soc. Faraday Trans. I 74, 1346 (1978).

Bowen, B. D. and Epstein, N., J. Colloid Interface Sci. 72, 81 (1979).

Bowen, B. D., Levine, S. and Epstein, N., J. Colloid Interface Sci. 54, 375 (1976).

Chari, K., Convective Transport of Colloidal Species Toward Substrates, Ph.D. Dissertation in progress, Rensselaer Polytechnic Institute, Troy, NY, 1984.

Chari, K. and Rajagopalan, R., Analytical Solutions to the Rate of Deposition of Brownian Particles in the Presence of Interaction Forces, Paper presented at the 58th ACS Colloid and Surface Science Symposium, Pittsburgh, PA, June 1984.

Clint, G. E., Clint, J. H., Corkill, J. M. and Walker, T., J. Colloid Interface Sci. 44, 121 (1973).

Dabros, T., Ph.D. Dissertation, Jagiellonian University, Krakow, Poland, 1977.

Dabros, T. and Adamczyk, Z., Chem. Eng. Sci. 34, 1041 (1979).

Dabros, T., Adamczyk, Z. and Czarnecki, J., J. Colloid Interface Sci. 62, 529 (1977).

Dabros, T. and van de Ven, T. G. M., A Direct Method for Studying Particle Deposition on Solid Surfaces, PGRL Report No. 265, Pulp and Paper Research Institute of Canada, Pointe Claire, Canada, 1982.

Dabros, T. and van de Ven, T. G. M., J. Colloid Interface Sci. 92, 403 (1983a).

Dabros, T. and van de Ven, T. G. M., J. Colloid Interface Sci. 93, 576 (1983b).

Dahneke, B., J. Colloid Interface Sci. 48, 520 (1974).

Dahneke, B., J. Colloid Interface Sci. 50, 89 (1975a).

Dahneke, B., J. Colloid Interface Sci. 50, 194 (1975b).

Delahay, P. (ed.), Advances in Electrochemistry and Electrochemical Engineering, Vol. 4, Wiley-Interscience, New York, NY, 1966.

Demchak, R. and Matijević, E., J. Colloid Interface Sci. 31, 257 (1969).

Gregory, J. and Wishart, J., Colloids Surfaces 1, 313 (1980).

Hansen, F. K. and Matijević, E., J. Chem. Soc. Faraday Trans. I 76, 1240 (1980).

Hermansson, H. P., Chemica Scripta 12, 102 (1977).

Hull, M. and Kitchener, J. A., Trans. Faraday Soc. 65, 3093 (1969).

Kallay, N. and Matijević, E., J. Colloid Interface Sci. 83, 289 (1981).

Kallay, N., Nelligan, J. D. and Matijević, E., J. Chem. Soc. Faraday Trans. I 79, 65 (1983).

Karis, T. E., Diffusional Mass Transfer to a Rotating Disk, M. S. Thesis, Rensselaer Polytechnic Institute, Troy, NY, 1979.

Kim, J. S. and Rajagopalan, R., Colloids Surfaces 4, 17 (1982).

Kolakowski, J. E. and Matijević, E., J. Chem. Soc. Faraday Trans. I 75, 65 (1979).

Kuo, R. J. and Matijević, E., J. Chem. Soc. Faraday Trans. I 75, 2014 (1979).

Kuo, R. J. and Matijević, E., J. Colloid Interface Sci. 78, 407 (1980).

Levich, V. G., Physicochemical Hydrodynamics, Prentice Hall, Englewood Cliffs, NJ, 1962.

Li, N. N. (ed.), Recent Developments in Separation Science, Vol. IV, CRC Press, West Palm Beach, FL, 1978.

Marshall, J. K. and Kitchener, J. A., J. Colloid Interface Sci. 22., 342 (1966).

Matijević, E., Pure Appl. Chem. 52, 1179 (1980).

Matijević, E., Accounts Chem. Res. 14, 22 (1981).

Matijević, E. and Scheiner, P., J. Colloid Interface Sci. 63, 509 (1978).

Matijević, E. and Wilhelmy D. M., J. Colloid Interface Sci. 86, 476 (1982).

Milić, N. B. and Matijević, E., J. Colloid Interface Sci. 85, 306 (1982).

Partch, R., Matijević, E., Hodgson, A. W. and Aiken, B. E., J. Polym. Sci.: Polym. Chem. Ed. 21, 961 (1983).

Pleskov, Y. and Filinovskii, J., The Rotating Disk Electrode, Consultants Bureau, New York, NY, 1976.

Poehlein, G. W., Ottewill, R. H. and Goodwin, J. W. (eds.), Science and Technology of Polymer Colloids, Vol. II, Martinus Nijhoff, The Hague, The Netherlands, 1983.

Prieve, D. C. and Carrera, J. P., pp. 23-39 in Becher and Yudenfreund (1978).

Prieve, D. C. and Lin, M. J., J. Colloid Interface Sci. 76, 32 (1980).

Prieve, D. C. and Ruckenstein, E., AIChE J. 20, 1178 (1974).

Prieve, D. C. and Ruckenstein, E., J. Colloid Interface Sci. 57, 547 (1976).

Rajagopalan, R. and Chu, R. Q., J. Colloid Interface Sci. 86, 299 (1982).

Rajagopalan, R. and Karis, T. E., pp. 73-81 in Bennett (1979).

Rajagopalan, R. and Kim, J. S., J. Colloid Interface Sci. 83, 428 (1981).

Rajagopalan, R. and Tien, C., Canad. J. Chem. Engr. 55, 246 (1977).

Rajagopalan, R. and Tien, C., pp. 179-269 in Wakeman (1979).

Riddiford, A. G., pp. 47-116 in Delahay (1966).

Ruckenstein, E., J. Colloid Interface Sci. 66, 531 (1978).

Ruckenstein, E. and Prieve, D. C., J. Chem. Soc. Faraday Trans. II 69, 1522 (1973).

Ruckenstein, E. and Prieve, D. C., pp. 107-137 in Beddow and Meloy (1980).

Sapieszko, R. S. and Matijević, E., J. Colloid Interface Sci. 74, 405 (1980).

Spielman, L. A. and Friedlander, S. K., J. Colloid Interface Sci. 46, 22 (1974).

Sugimoto, T. and Matijević, E., J. Colloid Interface Sci. 74, 227 (1980).

Tewari, P. H. and Campbell, A. B., pp. 83-92 in Li (1978).

Tien, C., Wang, C-S. and Barot, D. T., Science 196, 983 (1977).

van de Ven, T. G. M., Adv. Colloid Interface Sci. 17, 105 (1982).

van de Ven, T. G. M., Dabros, T. and Czarnecki, J., J. Colloid Interface Sci. 93, 580 (1983).

Visca, M. and Matijević, E., J. Colloid Interface Sci. 68, 308 (1979).

Wakeman, R. J. (ed.), Progress in Filtration and Separation, Vol. 1, Elsevier, Amsterdam, The Netherlands, 1979.

Wnek, W. J., The Role of Surface Phenomena and Colloid Chemistry in Deep Bed Liquid Filtration, Ph.D. Dissertation, Illinois Institute of Technology, Chicago, IL, 1973.

Wnek, W. J., Gidaspow, D. and Wasan, D. T., Chem. Eng. Sci. 30, 1035 (1975).

Wnek, W. J., Gidaspow, D. and Wasan, D. T., J. Colloid Interface Sci. 59, 1 (1977).

Zimmer, S. L. and Dahneke, B., J. Colloid Interface Sci. 54, 329 (1976).

REFERENCES: SECTION B. ADSORPTION OF COLLOIDAL PARTICLES IN PACKED BEDS

Aris, R. and Amundson, N. R., First-Order Partial Differential Equations with Applications, Prentice Hall, Englewood Cliffs, NJ, 1973.

Chowdiah, P., Wasan, D. T. and Gidaspow, D., AIChE J. 27, 975 (1981).

Dahneke, B., J. Colloid Interface Sci. 50, 89 (1975).

Heertjes, P. M. and Lerk, C. F., Trans. Inst. Chem. Eng. 45, T129 (1967).

Kallay, N. and Matijević, E., J. Colloid Interface Sci. 83, 289 (1981).

Kallay, N., Nelligan, J. D. and Matijević, E., J. Chem. Soc. Faraday Trans. I 79, 65 (1983).

Kolakowski, J. E. and Matijević, E., J. Chem. Soc. Faraday Trans. I 75, 65 (1979).

Kuo, R. J. and Matijević, E., J. Chem. Soc. Faraday Trans. I 75, 2014 (1979).

Kuo, R. J. and Matijević, E., J. Colloid Interface Sci. 78, 407 (1980).

Matijević, E. and Kallay, N., Croat. Chem. Acta 56, 649 (1983).

Matijević, E., Kuo, R. J. and Kolny, H., J. Colloid Interface Sci. 80, 94 (1981).

Nelligan, J. D., Kallay, N., and Matijević, E., J. Colloid Interface Sci. 89, 9 (1982).

Rajagopalan, R. and Chu, R. Q., J. Colloid Interface Sci. 86, 299 (1982).

Rajagopalan, R. and Tien, C., AIChE J. 22, 523 (1976).

Rajagopalan, R. and Tien, C., pp. 179-269 in Wakeman (1979).

Rajagopalan, R., Tien, C., Pfeffer, R. and Tardos, G., AIChE J. 28, 871 (1982).

Thompson, G., Kallay, N. and Matijević, E., Chem. Eng. Sci. 38, 1901 (1983).

Wakeman, R. J. (ed.), Progress in Filtration and Separation, Vol. 1, Elsevier, Amsterdam, The Netherlands, 1979.

Wnek, W. J., Gidaspow, D. and Wasan, D. T., Chem. Eng. Sci. 30, 1035 (1975).

Yao, K. M., Habibian, M. T. and O'Melia, R., Environ. Sci. Technol. 5, 1105 (1971).

PART III

CONCENTRATED DISPERSIONS

6

Static and Dynamic Structure of Interacting Dispersions

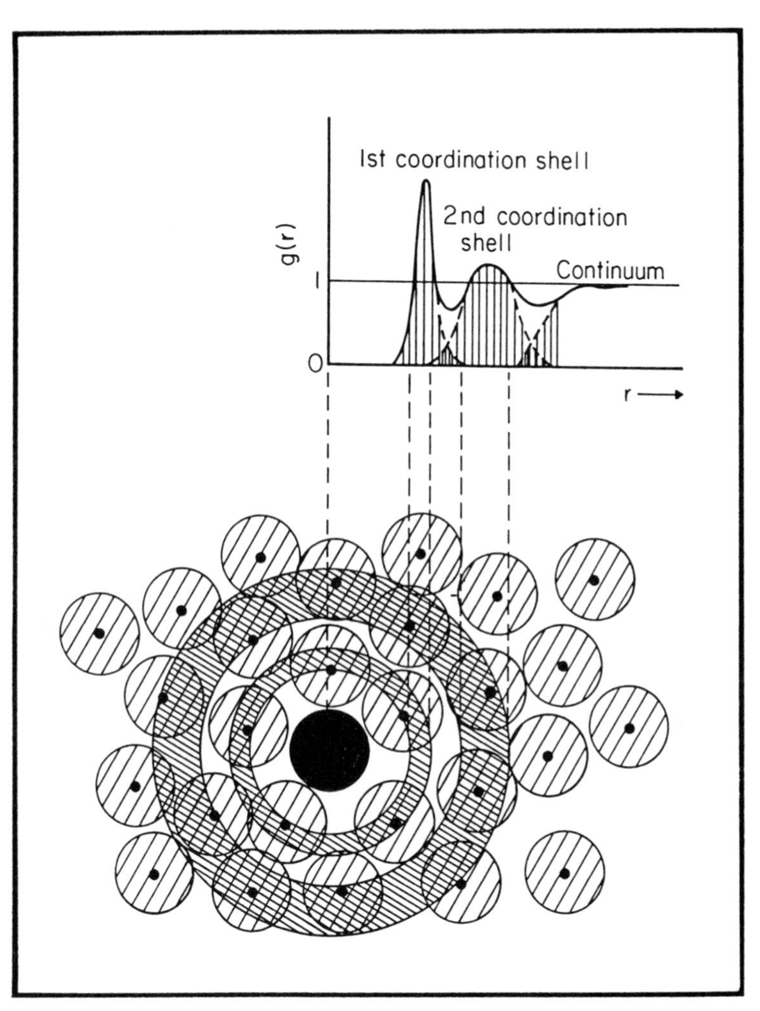

CHAPTER 6

STATIC AND DYNAMIC STRUCTURE OF INTERACTING DISPERSIONS

Overview

The phenomena and properties associated with concentrated dispersions differ markedly from those observed in dilute systems. While many of the qualitative and quantitative features of dilute dispersions can be interpreted exclusively in terms of pair interactions (i.e., in terms of interaction energies between two particles or between a particle and a surface), such a procedure is hardly adequate in the case of concentrated dispersions. Many notable examples of dispersions of practical interest such as printing inks and paints fall under the latter class and their macroscopically-observed properties such as color, opacity, gloss, consistency and texture depend on the stability and local structure of the dispersions. A word of caution is perhaps needed here concerning what is implied by the term 'concentrated'. A 'concentrated' dispersion is perhaps conveniently defined as one whose properties of interest are dependent on the concentration of the dispersion, although the actual volume fraction of the dispersed material may be extremely low (say, on the order of 0.01). A more fundamental definition is that the phenomena in a concentrated dispersion are dictated by <u>many-body</u> interactions.

To illustrate this, we note that the microscopic interactions that give rise to concentration dependence of macroscale properties are essentially of two kinds; the first, which has received more attention in the literature until recently, is hydrodynamic in origin, is typically long-ranged, and extends to many particle-radii from any given particle (this is discussed further in Chapter 7); the second principal contribution arises from intermolecular and coulombic forces among the particles. Since the length scale over which the coulombic (or, electrostatic) forces are important increases with decreasing electrolyte concentration, the delineation of particle concentration as 'high' or 'low' is relative and depends on the electrolyte concentration in the liquid and the nature and concentration of charges on the particles. A striking consequence of this is that the effect of many-body interactions can be felt at particle concentrations on the order of 0.5% (by volume) or lower. Efremov (1976) discusses in great length the role of colloidal forces in many dispersions of industrial and biological importance and presents photographic illustrations of some cases in which many-body interactions lead to the formation of gels and crystalline structures in dispersions which otherwise would be considered 'dilute' [see Efremov (1976) in the reference section for Section A at the end of this Chapter].

Investigation of colloidal effects in such systems can be approached purely from the point-of-view of observed macroscopic <u>properties</u>, either dynamic (e.g., measurement of diffusion coefficients or rheological data) or static (e.g., osmotic pressures). However, interpretation of these measurements requires information on the <u>microstructure</u> of the dispersions. In the absence of experimental information on the latter (and without the benefit of appropriate many-body theories), focusing on macroscopic measurements leads to no more than empirical equations for the specific system studied (e.g., an empirical equation of state for the osmotic pressure). A systematic analysis of concentration effects thus requires information on both the equilibrium structure <u>and</u> the dynamic structure -- obtained preferably by non-intrusive techniques.

This Chapter reviews the research in this area, but before proceeding further, we note that research on many-body interactions in colloidal dispersions has grown very rapidly in recent years. The contributions come from a highly interdisciplinary group of researchers ranging from theoretical chemists, physicists and biologists to chemical and materials engineers; correspondingly the goals and the tools differ markedly. Nevertheless, the general thrust of these can be grouped under the three following categories:

1. Experimental studies of static and dynamic structures in colloidal systems and phenomenological and analytical interpretations of such structures to study their microscale origin.

2. More sophisticated statistical mechanical descriptions of colloids as supramolecular fluids in order to deduce macroscale phenomena and properties.

3. Use of colloids as model many-body systems to investigate general theories of many-body interactions in molecular as well as supramolecular systems.

The purpose of this Chapter is to discuss developments in theoretical and experimental techniques for probing many-body interactions in general colloidal systems. We shall restrict the discussions to methods specifically aimed at the description of the microscale structure in terms of the interactions between the particles. In addition, recent developments in experimental techniques (particularly, non-intrusive, radiation scattering techniques) will be summarized along with how these can be used to extract information on electrochemical and surface chemical parameters of the system and the interaction forces. However, although the discussion of the dynamic aspects of the microscale structure (i.e., the evolution of the local structure from one moment to the next) cannot be separated from the associated diffusion properties, the bulk of the theoretical considerations that go into the derivation of diffusion coefficients in terms of hydrodynamic and colloidal interactions will be reserved for the next Chapter, on Transport Properties.

The review here will be divided as follows:

A. Static Structure: Colloidal Crystals,

B. Static Structure: Colloidal Fluids,

C. Dynamic Structure, and

D. Experimental Techniques.

A. Static Structure: Colloidal Crystals

The crystal-like structures observed in monodispersed colloidal species have attracted considerable attention in recent years for a variety of reasons. Although the origin of this interest, particularly in colloid science, can be traced to the initial experimental reports of Hiltner and Krieger (1969), Vanderhoff et al. (1970) and Krieger and Hiltner (1971) on iridescence in latex dispersions due to the Bragg scattering by the ordered state, the formation of order in supramolecular systems has been recorded and studied for over forty years. For instance, the unusual mechanical properties of bentonite sols and the optical properties of Schiller layers in iron oxide sols have been traced to the formation of order in these dispersions of nearly-monodispersed particles [Langmuir (1938); see also the recent work of Maeda and Hachisu (1983) on optical observations of Schiller layers]. The biological literature offers a fascinating variety of ordered structures of colloidal dimensions (see Figures 6.A.1 and 2); among these are the body-centered cubic (b.c.c.) structures observed in Bushy Stunt Virus [Bernal and Fankuchen (1941)] and the face-centered cubic structures observed in solutions of Tipula Virus [commonly known as Tipula Iridescent Virus, because of the opal-like iridescence of the crystals; Williams and Smith (1957)]. Many additional references on ordering in natural and man-made materials may be found in the numerous recent publications cited in the following pages of this report. The fact that many of the problems encountered in biology and medicine on self-assembly of microscopic species are closely akin to the classical problems in order-disorder transitions in chemistry and physics has led to an additional growth in activities in experimental and theoretical studies of supramolecular systems [see, for example, Thomsen (1982)], and it will be

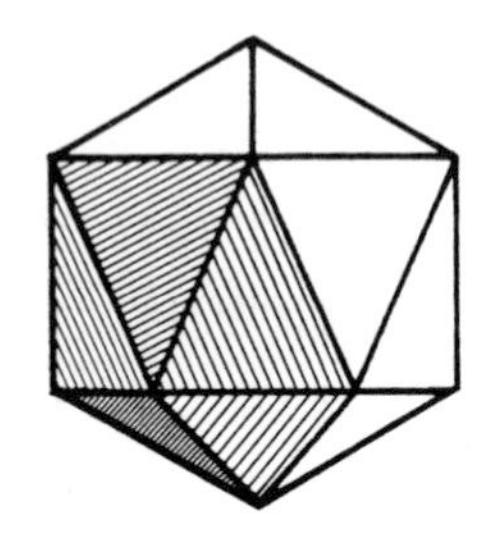

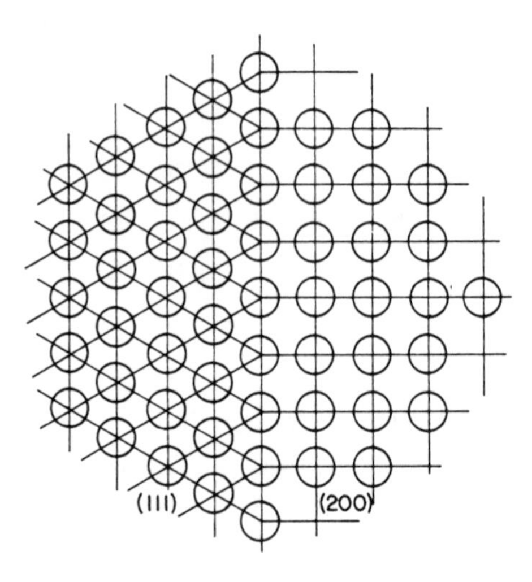

Figure 6.A.1 Tipula Iridescent Virus [from Pieranski (1983)].

This large insect virus has an icosahedral structure and has a diameter of about 1300 Å. The crystals formed have a face-centered cubic structure. The (111) and (200) Miller planes are shown above.

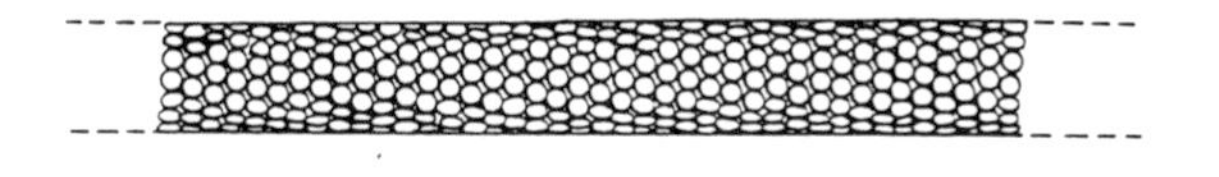

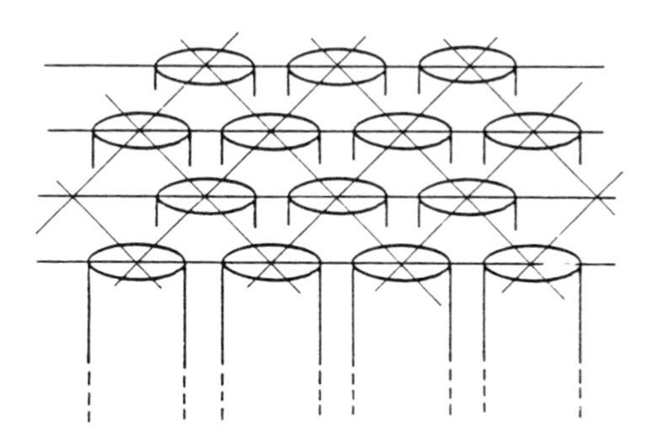

Figure 6.A.2 Tobacco Mosaic Virus [see Pieranski (1983)].

The rod-shaped TMV is about 150-170 Å in diameter and about 3000 Å in length. It is built of identical copies of protein macromolecules arranged in a helical structure (top Figure). The bottom Figure is a sketch of two-dimensional crystalline order formed by TMV.

fruitless to review every aspect of this area here. Instead, we shall confine our attention in the following discussion to those experimental and theoretical studies that focus on model colloidal systems and on using such systems to study colloidal interactions. Nevertheless, numerous references to reports of related interest (dealing with systems of only peripheral importance here) can be found in the many papers and reviews cited in this Report. For convenience, the present review of research activities on colloidal crystals will be divided into three, somewhat overlapping, parts:

1. Ordering in Colloids: Implications to Interaction Forces,
2. Macroscale Properties of Ordered Colloids, and
3. Colloidal Crystals as Model Many-Body Systems.

A.1 Ordering in Colloids: Implications to Interaction Forces

An extensive review of experimental observations on ordering in a large class of colloidal systems (including ionic micelles, macroions, proteins and virus particles) has been presented recently by Ise and Okubo (1980). The choice of the experimental technique for observing ordering depends on the dimensions of the colloidal material and may include ultramicroscopy or radiation scattering techniques such as light scattering, X-ray scattering and neutron scattering. The results on latex dispersions have been most striking and convincing in view of the fact that the iridescence due to ordering can be observed with the naked eye and the ordered structures themselves can be seen directly through ultramicroscopes [see, for example, Hachisu, Kobayashi and Kose (1973); Hachisu and Kobayashi (1974); Hachisu et al. (1976); Kose et al. (1973); Kose and Hachisu (1974)]. While these latter reports could establish the presence of ordered structures only in the layers closest to the objectives (of the microscope) or near the walls of the containers, an ingenious series of experiments using a mixture of latex particles and grains of ion-exchange resins has been performed recently by Ise et al. (1982) to demonstrate that stable, ordered structures exist throughout the suspension. Ise et al. report that the resin particles (anionic or cationic) were found to be trapped all through the dispersions in a face-centered cubic lattice of the latex crystals and were found to remain stable for over nine months in some cases. The resin particles could be made to settle to the bottom by changing the ionic concentration; in addition, the ordering and the trapping of the resin particles were found to be reversible with changes in the temperature. Although ordering in the presence of ion-exchange resins is likely to be different from that in their absence, the report of Ise et al. illustrates many of the features one might expect regarding the formation of order in supramolecular systems.

As emphasized by the title of this section, the implication of this phenomenon to the nature of interaction forces in concentrated dispersions has received considerable attention in the literature and, naturally, the examination of this question goes hand in hand with attempts to explain the ordering in terms of the interaction forces. For ease of further discussion, one may conveniently classify the past and present research directions in this area in the following manner.

(i) Effective Hard-Sphere Interaction

The simplest possible analysis of crystallization is to treat the dispersion as a suspension of hard, mutually-impenetrable spheres. This is of course equivalent to assuming the simplest possible pair interaction between the particles, namely, a hard-sphere interaction. The adequacy of this approximation, some of the needed (empirical and theoretical) modifications of this approximation, and the implication of this approach to the classical theories of phase transitions have been discussed extensively in the literature. These will be reviewed below briefly since there is sufficient merit in the approaches based on hard-sphere models under certain conditions and since, in addition, hard-sphere approximations frequently form a good reference point for subsequent refinements to accommodate softer repulsions or attractive forces. The latter advantage is particularly relevant in the case of fluid-like colloidal structures discussed in Section B of this Chapter.

(ii) Analogies with Electron Crystals and Strong Electrolytes

Attempts to explain ordering and its implications to the nature of interaction forces have also been made from the point of view of known systems such as electron crystals and solutions of strong electrolytes. Although such analogies are only approximate, they have been helpful in providing some additional insight into the nature of many-body interactions relevant in both molecular and supramolecular systems. These developments are also summarized in the following pages.

(iii) Reexamination of Traditional Interaction Potentials

The inability of traditional Derjaguin-Landau-Verwey-Overbeek theory (see Chapter 4) of colloidal interaction potentials to explain some of the experimental observations concerning the structure of the colloidal crystals suggests that a reexamination of assumptions involved in the DLVO derivation of potentials may be necessary. There have been some notable advances in this direction in the last two years. These and some historical papers of related interest are also discussed below.

It was pointed out by Kirkwood (1939) over forty years ago that a system of particles interacting via hard-sphere potentials would undergo a phase transition from a disordered, fluid-like structure to an ordered, solid-like structure at a volume fraction well below the one corresponding to the closest packing. The molecular dynamics 'experiments' of Alder and Wainwright (1962) eventually established the validity of Kirkwood's prediction, and subsequent statistical mechanical extensions and refinements have placed the range of volume fractions where the two phases coexist at roughly 0.5 to 0.55 [see, for example, Ziman (1979)]. This hard-sphere phase transition is generally known as the Kirkwood-Alder transition and offers, at least in retrospect, a conceptual basis for understanding phase transitions in colloids. First suggested by Stigter (1954), this correspondence between colloidal phase transitions and Kirkwood-Alder transition was resurrected about twenty years later by Wadati and Toda (1972) [see, also, van Megen and Snook (1975a,b)]. As mentioned earlier, using this correspondence to understand formation of order in colloids is equivalent to assuming the pair interaction between the particles to be of the hard-sphere type -- a simplification that is acceptable in the case of uncharged particles but which is only an approximation even in the case of sterically-stabilized particles. The colloidal forces modify the hard-sphere interaction in two ways: first, the repulsive interaction between the overlapping electrical double layers increases the 'effective' diameter of the particles; secondly, since the electrostatic repulsion decays over a finite region and is reduced in intensity by the attractive force, the combined forces soften the interaction potential. Many of the subsequent attempts on casting colloidal order/disorder transformations in terms of Kirkwood-Alder transitions have been concerned with devising suitable methods for defining effective hard-sphere diameters in terms of empirical observations or known pair interaction potentials. (This approach is, in fact, implicit in the Wadati-Toda proposal.) Among these are the papers of Brenner (1976), Barnes et al. (1978), Hachisu and Takano (1982), and Furusawa and Yamashita (1982), which specify the effective hard-sphere diameter in terms of either the double layer thickness or the radial distance at which the pair potential has a preselected value (usually of the order of thermal energy). These methods are purely *ad hoc* procedures, and although the dependence of the effective hard-sphere diameter on electrolyte strength and other electrochemical parameters can be artificially included in this manner, they provide no dependable criterion for selecting the magnitude of the effective hard-sphere diameter. For instance, the procedure suggested by Barnes et al. (1978) requires experimental values of volume fractions at which a completely-ordered dispersion 'melts'; the effective diameters are then calculated on the assumption that the effective volume fraction at melting is 0.74 (i.e., volume fraction corresponding to closest packing in a face-centered cubic lattice) and the 'freezing point', i.e., the volume fraction at which a disordered dispersion begins to solidify, is estimated on the assumption that the ratio of the melting point to the freezing point is the same as the Kirkwood-Alder ratio of 1.1 (i.e., 0.55/0.5). There is of course no guarantee that this estimate would be the same as the experimentally-observed freezing point. Since there is no reason to expect the existence of a unique value for the effective diameter, an extensive discussion of these procedures is perhaps unnecessary; nevertheless, the empirical and theoretical experience gained from these and similar approaches can, in fact, be seen to provide a convincing argument for the need for statistical mechanical treatments of the problem. To this end, the following observations, based on the above papers and other (somewhat more rigorous) approaches, will be useful.

(i) If the hard core of the potential is fairly steep and the influence of attraction is not strong enough to cause flocculation, an ad hoc procedure of the above type is adequate for estimating the volume fractions at which freezing and melting occur (see Figure 6.A.3). In fact, the dependence of the Debye length on particle concentration also can be accommodated in such an *ad hoc* approach; see, for example, Barnes et al. (1978).

(ii) However, the other properties (such as osmotic pressures or isothermal compressibilities) derived from hard-sphere fluids using the above effective diameters may not necessarily match the actual values; that is, there is no uniquely defined value for the effective hard-sphere diameter that can represent the dispersion in its totality.

(iii) Consequently, these ad hoc procedures must be seen as highly-simplified exploratory attempts to represent the many-body effects in terms of convenient reference potentials. In fact, as will be

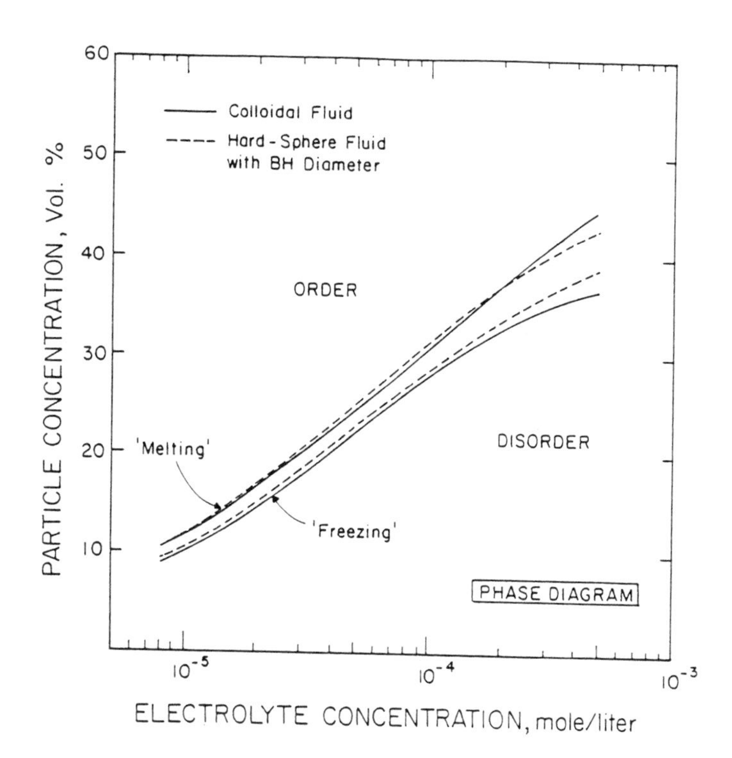

Figure 6.A.3 Phase diagram for a colloidal fluid and that for an equivalent hard-sphere fluid [from Castillo, Rajagopalan and Hirtzel (1984)].

The hard-sphere diameter is obtained from the Barker-Henderson perturbation theory.

discussed in Section B of this Chapter, hard-sphere fluids are frequently used as reference fluids in perturbation theories; therefore, the above procedures may be thought of as zeroth-order perturbation approaches, although selecting the effective diameter on the basis of the magnitude of the pair potential at that distance does not rigorously account for the statistical nature of many-body interactions.

(iv) There is no need for the existence of a significant secondary minimum in the interaction potential for the ordering to occur. (It was believed originally that the formation of periodic colloid structures required a significantly deep secondary minimum in the pair potential.)

Effective Hard-Sphere Interactions

The last two points above have been stressed by Israelachvili and Ninham (1977) in their review of the historical development of the theory of intermolecular forces. Israelachvili and Ninham have also emphasized the need for proper statistical mechanical treatments to fully understand the stability and the long- and short-ranged structures observed in colloidal, micellar and biological systems. These are carried a little further in the paper of Forsyth et al. (1978), in which Onsager's classical work on ordering in nematic fluids [Onsager (1949)] is used to explain ordering in solutions of Tobacco Mosaic Virus. Forsyth et al. (1978) also examined the usefulness of Lindemann's empirical rule for melting of crystals, which states that an ordered structure will melt when the root mean square displacement of the particles from their lattice sites is a characteristic fraction of the lattice spacing. Although this is also an approximate procedure, it serves to demonstrate that the order/disorder transition in colloids is first-order. Maximization of entropy determines the conditions for order formation in a dispersion of hard rods or spheres confined to a finite volume. On the other hand, for dispersions in which interaction energy includes attraction and repulsion, minimization of the free energy determines the conditions under which a phase transition can be expected. This has been examined by Krumrine and Vanderhoff (1980) using a 12-particle icosahedron arrangement for the local structure of a colloid interacting via a DLVO potential. Both this and another 38-particle construction lead to a good reproduction of the experimental phase diagram, somewhat similar to the ones based on Kirkwood-Alder transition with effective hard-sphere diameters. While the analyses of Forsyth et al. (1978) and Krumrine and Vanderhoff (1980) do not add any significant improvements to the earlier procedures, they indicate the increasing recognition of the importance of many-body interactions implicit in the order formation. This is evident from the increasing use of statistical mechanical ideas for describing structural transitions in dispersions [e.g., Snook and van Megen (1976a,b); van Megen and Snook (1975a,b, 1976, 1977)]. Snook and van Megen (1976a) and van Megen and Snook (1975a,b) discuss order formation from the point of view of effective hard-sphere approaches discussed above; these general ideas have been already outlined in the above discussion. The other papers are more general and deal with the disordered state as well; these will be discussed in Sections A.2 and B of this Chapter. Some additional discussions of these and other papers are presented in Castillo (1982), Castillo, Rajagopalan and Hirtzel (1984), and Rajagopalan and Hirtzel (1984). Finally, we note that more general studies of structural transitions in colloids are becoming available in the literature. Among these are Chaikin et al. (1982), Joyce (1983) and Sogami et al. (1984). Chaikin et al. demonstrate, using free energy computations as functions of temperature and ionic strength, that melting and re-entrant transitions (i.e., transition back to the original structure) are possible with monotonic temperature changes when the temperature dependence of the dielectric constant is included in the computations. Joyce (1983) uses a Lennard-Jones/Devonshire-type cell model to examine what type of crystal structure will be favored under any given condition (see Figures 6.A.6 and 7 in Section A.2). Similarly, Sogami et al. (1984) use lattice sums to show that the b.c.c. structure is favored in very dilute dispersions of macroions. All these studies are restricted to monodispersed colloids and it is too early at this time to draw definitive conclusions concerning their implications with respect to the nature of interaction forces.

Analogies With Wigner Crystals and Strong Electrolytes

As mentioned earlier, incorporation of many-body effects has also been approached through analogy with known many-body systems such as electron crystals and strong electrolytes so that the experience gained regarding the nature of interaction forces in these latter cases can be used to understand colloidal dispersions. Three such examples will be reviewed here. A simple approach that can be examined easily is one in which a crystal formed by highly charged macroions is assumed to be analogous to structures found in metals and electron crystals (known as Wigner crystals). In these latter cases, charges of one sign are localized in a lattice structure while the charges of opposite signs are delocalized and form the neutralizing background. This analogy is, of course, superficial since the nature of delocalization is fundamentally different in the latter cases (i.e., quantal in metals, whereas it is thermal or entropic in colloids). Nevertheless, these offer convenient working models because of the extensive literature available on them [see, for example, March and Tosi (1974); Care and March (1975)]. If the background can be assumed to be uniform and structureless, the electrostatic interactions in metals and charged macroions may be taken to be the same and one can then use different lattice structures to determine which would be more stable [Pieranski (1983)]. The electrostatic energy in these cases, however, diverges inversely with the radius of the cell in which each particle is confined (known as the Wigner-Seitz cell). To make this approach more accurate the entropic contributions of the particles as well as the counterions in the background must be included. A somewhat different, and more rigorous, approach using Wigner lattices has been tried by Marcelja, Mitchell and Ninham (1976). The particle is assumed to be confined to a spherical region and the potential distribution is obtained from the solution of the Poisson-Boltzmann equation subject to the constraint that the degree of dissociation of surface charge groups is determined self-consistently. Attractive interactions can be included in the net interaction energy expressions. Marcelja, Mitchell and Ninham use this model to construct the phase diagram by resorting to the empirical Lindemann rule for melting [Lindemann (1910)] mentioned earlier. It is known from Monte Carlo studies on Wigner lattices that melting occurs when the root mean square displacement of the particles exceeds about 0.08 times the lattice spacing. In fact, for potentials of the form r^{-n} (for $n \geqslant 4$), the fraction of lattice spacing needed is about 0.1. Therefore, whether the interaction potential is very soft (unscreened coulomb form) or hard (r^{-n}; $n \gtrsim 10$), at least qualitatively correct phase diagrams can be constructed. The computations of Marcelja, Mitchell and Ninham (1976) are in good agreement with the experimental data of Hachisu, Kobayashi and Kose (1973) at low particle concentrations; however, the fluctuations in experimental data in this region are rather large. The Wigner lattice model deviates significantly from experimental observations for volume fractions larger than about 0.2. This is perhaps not surprising in view of the role of counterions on the net interaction forces in concentrated dispersions and will be discussed further in the last segment of this section (Section A.1).

In contrast to the above, semiempirical analogies, comparing dispersions of charged macroions or particles with strong electrolytes is more justifiable and meaningful from the statistical mechanical point of view. This was recognized by Langmuir (1938) over forty years ago as mentioned earlier [see, also, Pieranski (1983)]. However, although Langmuir's early analysis deserves credit for considering the problem from a very general point of view, the description of the interaction potentials in terms of the Debye-Hückel theory (see Chapter 3, Section A) leaves his theory open to some serious criticisms:

(1) For instance, the linearization of the Poisson-Boltzmann equation fails at large particle concentrations and for highly charged particles.

(2) The mean field approach characteristic of the Debye-Hückel theory excludes all correlations between ions in solutions other than those which describe the exponential screening of each ion by all the others.

These objections can be circumvented by considering oscillatory screening that occurs at large concentrations of macroions [Kirkwood (1936)]. When the dimension of the particle or macroion is of the order of the Debye length, the screening of repulsion in a concentrated dispersion can be oscillatory and can lock the particles into a lattice. This concept has been explored by Hastings (1978a) in an attempt to understand the discrepancy between some calculations based on exponential screening of the potential. For instance, the previously discussed descriptions of colloid crystals on the basis of minimization of free energy use the screened version of coulomb repulsion to compute the energy of interaction. The crystalline phase in the case of pure repulsion is dictated by the minimization of repulsion (in the absence of entropic effects). This 'coulomb solid model' has been used by Williams, Crandall and Wojtowics (1976) to predict the melting temperature as a function of particle concentration, and by Hastings (1978b) to explain the yield stress observed in solutions of Turnip Yellow Mosaic Virus at room temperatures. However, these models were semiempirical since one of the parameters in the model, which is a function of the Debye length, had to be determined from experimental data. The Debye lengths determined in this manner are considerably higher than the actual (estimated) values (which were about a few Angstroms). Hastings (1978a) seeks to avoid this discrepancy by considering the stratification of ions (which in reality are of _finite_ size and are not point charges) at high electrolyte concentrations. Kirkwood's theory predicts a mean field transition to a phase in which the interion potential oscillates with a wavelength that is only weakly dependent on electrolyte concentration. This critical wavelength is interpreted by Hastings as being roughly equal to the lattice constant in the crystalline phase (Figure 6.A.4). The calculations of Hastings, which can be used to estimate the minimum charge required for crystallization for any given particle size, lead to qualitative agreement with experiments. A longer discussion of this theory is given in Pieranski (1983); Hastings himself suggests some possible improvements of the theory [Hastings (1978a)]; however, more experimental information on ordered structures using well-defined model colloids is needed before the available theories can be examined more closely and useful improvements can be identified.

Reexamination of Traditional Interaction Potentials

Somewhat in parallel with the above developments, a closer examination of the effect of counterions in concentrated dispersions of macroions or particles has also been undertaken by some investigators [see Ise (1983)]. Ise (1983) reports that the experimentally-observed interparticle spacing

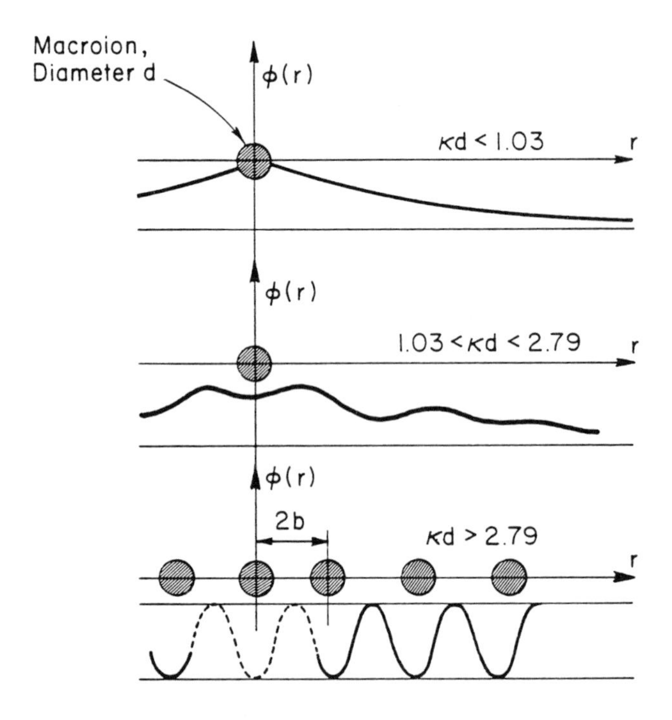

Figure 6.A.4 Different types of screening in Kirkwood-Hastings theory of strong electrolytes [from Pieranski (1983)].

The inverse screening length is denoted by κ.

increases with increasing salt concentration and decreases with increasing number of charges on the particles [see, also, Ise and Okubo (1980); Ise et al. (1980, 1983a,b)]. These observations cannot be explained in terms of the traditional DLVO interaction potentials, whose behavior with respect to increasing electrolyte strength or increasing surface charge on the particles will dictate changes in interparticle spacings that are just the opposite of what is observed. Ise's explanation of these observations attributes the reason for these to the intermediary role played by the counterions on the interactions between the macroions (see Figure 6.A.5). As sketched in Figure 6.A.5, the presence of counterions in the neighborhood of the macroion induces an attractive force between the macroion and the counterions. The component of this attractive force along the line connecting the centers of the macroions contributes to the reduction in the repulsion between the macroions [see, also, Feynman, Leighton and Sands (1963)]. The general thrust of these arguments may be summarized as follows.

(i) The small-angle X-ray scattering (SAXS) data and the small-angle neutron scattering (SANS) data on the experimentally-observed interparticle spacings at low concentrations indicate that the actual interparticle spacings are smaller than what would be expected if complete ordering were imposed by the restricted volume of the container. (This would be the case in a purely-repulsive system.) This observation and the fact that two different relaxation models have been observed in the intensity autocorrelation functions in dynamic light scattering experiments are taken to imply that the dispersion has both ordered and disordered regions [see Ise (1983) and Ise and Okubo (1980)].

(ii) Since a purely-repulsive system would form an ordered structure throughout the container, the average interparticle spacing must be determined by the density of the system. However, since the actual values of the spacing are lower than the above limit, attractive forces must be operative and must influence the ordering process. This attraction, which is traced to the intermediary role of counterions by Ise and co-workers, is therefore coulombic in origin.

(iii) As the charge of the particles increases, the strength of this attraction increases thereby reducing the interparticle spacing.

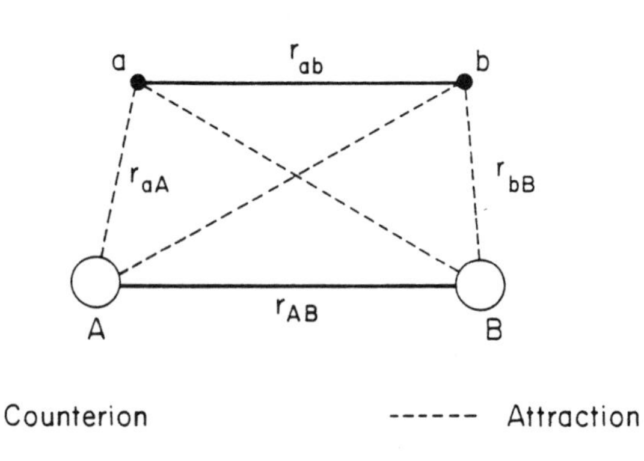

Figure 6.A.5 The origin of coulombic attraction in interacting dispersions [from Ise (1983)].

The attraction between a and A and between b and B can reduce the repulsion between A and B.

(iv) As simple salts are added to the dispersions, the influence of counterions is weakened by the screening caused by the salt. This, in turn, reduces the coulombic attraction and gives rise to an increase in the interparticle spacing.

Additional details on these arguments and other related references are given in the articles cited previously. Note that the last two points listed above [points (iii) and (iv)] cannot be explained using the traditional DLVO theory. Interestingly, Ise (1983) notes that Langmuir had pointed out in 1938 that the exclusion of coulombic attraction was one of the fallacies of colloid science. Recently, Sogami (1983) has rederived interaction potentials between charged particles by calculating the Gibbs free energy of the dispersion. Sogami states that the inclusion of the compressibility term (i.e., pressure times volume) in the calculation (i.e., using Gibbs energy instead of Helmholtz energy) amounts to including the coulombic attraction appropriately. Sogami's calculations show that the magnitude of the secondary minimum in the interaction potential first increases and then decreases as the electrolyte concentration increases, indicating that the strength of attraction increases first because of the counterion effect and subsequently decreases because of the screening of the counterions. Sogami has also presented computed values of the position of the secondary minimum as a function of electrolyte strength which agree well with the experimental interparticle distance, leading to the conclusion that ordering can be attributed to the secondary minimum in the interaction potential.

Finally, we note that there are other theories that predict the presence of coulombic attraction [Levine (1939); Kirkwood and Mazur (1952); Imai (1983a,b)]. For instance, Imai argues that coexistence of ordered and disordered phases cannot be explained solely in terms of minimization of free energies and claims that the difference in the relaxation times for the motion of particles on either side of the crystal/fluid interface must be considered.

The purpose of this Section (Section A.1) has been to summarize the current attempts to explain the formation of crystalline order and what it reveals about the nature of many-body interactions in dispersions. Although a good qualitative and some quantitative understanding of these exist now, discriminating between the available theories and determining what improvements are needed require additional experimental data.

A.2 Macroscale Properties of Ordered Colloids

As mentioned in the beginning of this Chapter, the equilibrium and transport properties of strongly-interacting dispersions are significantly different from what they are when many-body effects are negligible. Some of the major transport properties, for example, are discussed in Chapter 7. The central issue in all of these is the way in which the many-body interactions must be included in developing theories for the properties. While the onset of order leads to more complex behavior (e.g., appearance of yield stress), ordering also simplifies the analysis to some extent since the difficult problem of describing both the microstructure and many-body effects becomes simpler because of the more easily characterizable crystalline structure. Nevertheless, it is evident that, unless many of the problems discussed in Section A.1 (on the nature of many-body interactions) are resolved in an acceptable manner, theories of macroscale properties cannot be evaluated adequately. Despite this restriction, some attempts have been made in the literature to develop theories for predicting both equilibrium and transport properties of ordered dispersions. We review here briefly a few major advances on this aspect of the study of ordered colloids. This section will focus only on static properties (e.g., osmotic pressure, isothermal compressibilities, etc.), although occasional references to dynamic properties such as yield stresses and viscosities will be made where appropriate. These latter properties are discussed more fully in Chapter 7, Section C.

Most of the existing theories for equilibrium (as well as for transport) properties assume pairwise additivity of interaction potentials to account for the many-body effects. Then, for any given lattice structure, the free energies, osmotic pressures and other equilibrium properties can be computed using lattice theories that have been used in liquid state physics [see Barker (1963)]. The lattice theories assume that a central particle is confined to a cell consisting of many spherical shells, each of which accommodates a certain number of particles dictated by the lattice structure. (In principle, the shells do not have to be spherical, but this assumption simplifies the computations considerably.) One then derives an expression for the partition function from which the free energy (Gibbs or Helmholtz) can be obtained using standard definitions; other properties also follow from well-known relations [for example, the pressure is obtained from the negative derivative of Helmholtz free energy with respect to volume; see Münster (1974)]. There is no need to assume that the particles are fixed in the lattice sites; one can also consider more general cases in which the neighbors of the central particle are 'smeared' uniformly in their shells (the Lennard-Jones/Devonshire-type cell models; see Figures 6.A.6 and 7). This approach has been used by van Megen and Snook (1976, 1977, 1978), Castillo and Rajagopalan (1983), and Hirtzel and Rajagopalan (1983) to construct phase diagrams and osmotic pressures (see, for example, Figures 6.A.3 and 8). The osmotic pressures computed in this manner are in reasonable agreement with 'exact' Monte Carlo computations, but further tests are necessary before the usefulness of cell models can be evaluated accurately. Hirtzel and Rajagopalan (1983) show, using these calculations, that osmotic pressures estimated according to conventional hard-sphere equations of state may be in serious error compared to the actual pressures. Further, as discussed in Section A.1, order formation can also occur at much lower volume fractions than predicted by hard-sphere theories. (This is of significance in practical applications such as membrane filtration, where such 'gelation' eventually leads to fouling of the membrane.) While the papers cited above use standard DLVO expressions for the pair interaction potentials, the technique itself is obviously not restricted to any particular pair potential; in fact, the paper of Sogami et al. (1984) cited earlier uses the potential derived by Sogami (1983) to compute free energies and internal energies for various types of crystal structures. These computations can be extended to other equilibrium properties as well. Similarly, the works of Joyce (1983) and Chaikin et al. (1982), cited in Section A.1, also fall under the above type of cell models.

To avoid the assumption of pairwise additivity of potentials, Russel and Benzing (1981) use a 'self-consistent field' approach, which also starts with a central particle confined to a spherical cell. The sphere is assumed to have electrical charges uniformly distributed over its surface, and the fluid in the sphere (assumed to be Newtonian) contains counterions (equal in number to the charge on the particle) and any excess electrolyte. This spherical cell is embedded in a continuum which is assumed to have the properties of the bulk suspension. Any field (flow or electrical) imposed on the suspension is transmitted to the cell-suspension interface, thereby disturbing its spherical symmetry. Then the governing equations for the field are solved along with

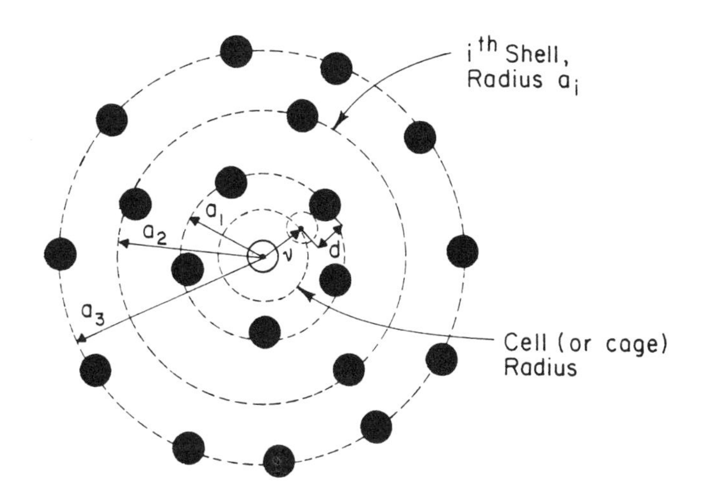

Figure 6.A.6 Schematic representation of one of the many possible cell models.

See text for details. See, also, Castillo, Rajagopalan and Hirtzel (1984).

CALCULATION OF THE CELL FIELD
IN THE SMEARING APPROXIMATION

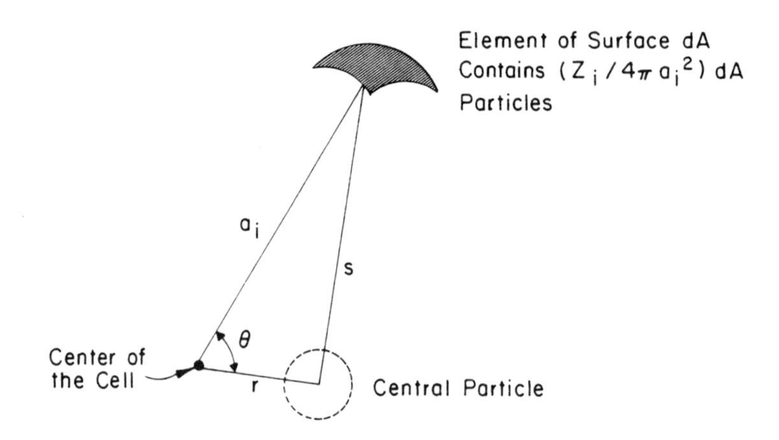

a_i : Distance of the ith Shell

Z_i : Number of Neighbors in the ith Shell

Ψ_i : Cell Field due to i^{th} Shell

$$\Psi_i(r) = \frac{Z_i}{2} \int_0^{\pi} u(s) \sin\theta \, d\theta$$

$$s = (a_i^2 + r^2 - 2a_i \, r \cos\theta)^{1/2}$$

Figure 6.A.7 Schematic representation for the calculation of the cell field in the model sketched in Figure 6.A.6 [see Castillo, Rajagopalan and Hirtzel (1984)].

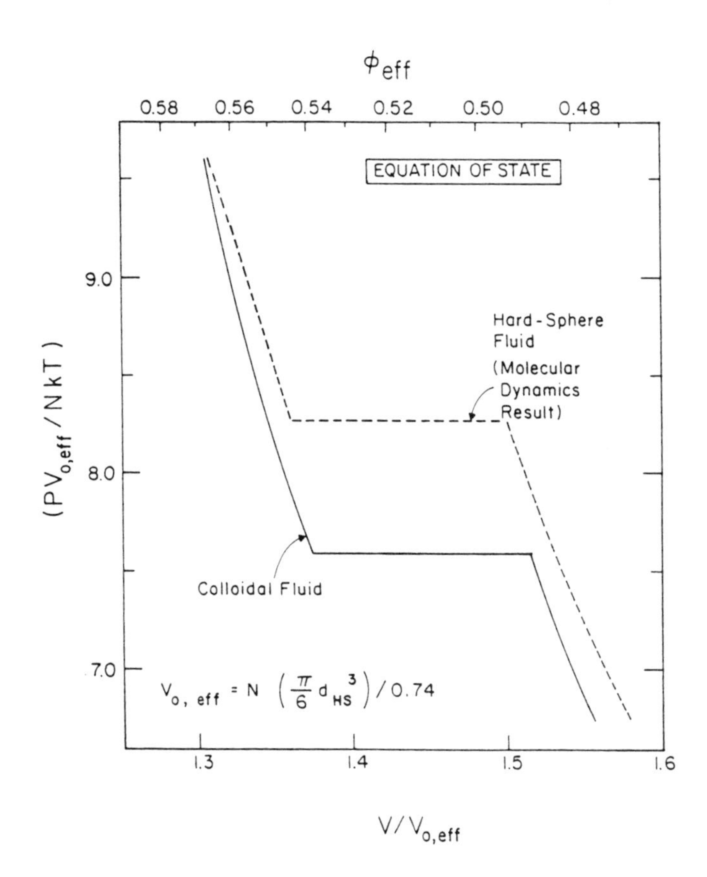

Figure 6.A.8 Equation of state computed from the cell model for the colloidal crystal and the Barker-Henderson perturbation theory for the liquid-like phase [see Castillo (1984) and Hirtzel and Rajagopalan (1983)].

appropriate boundary conditions, under suitable simplifications, and the unknown properties of the bulk are obtained using the 'self-consistency' condition, namely, that the averages within the cell must equal those in the surrounding bulk. The above summary outlines only the general features of this approach so that its differences from the previous cell models can be appreciated. Russel and Benzing (1981) have used this approach to compute the osmotic pressure, dielectric permittivity, dynamic viscosity and shear modulus as functions of volume fraction and electrochemical properties. Their results for the osmotic pressure follow the same qualitative trend predicted by the earlier approaches, but no quantitative comparisons have been made so far to study the relative merits of one approach over the other. Benzing and Russel (1981), however, have compared their predictions with experimental results obtained for polystyrene latex dispersions (up to about 25% by volume) and find qualitative agreement. Their results on viscosity and shear modulus are discussed in Chapter 7, Section C. Some additional references of interest (in the context of cell models) are discussed in Section B.3 of this Chapter and in Castillo, Rajagopalan and Hirtzel (1984).

Finally, it is useful to draw attention to a recent derivation of expressions for osmotic pressures by Grimson, who uses perturbation theory for the small-angle structure factor. Since the primary focus of this theory is on the disordered state, this work will be discussed in Section B of this Chapter. Grimson, however, uses this theory to predict conditions under which order formation and coexistence of the two phases can be expected. This theory is not expected to be accurate because of the approximate forms of the electrostatic and van der Waals potentials used. Although the theory predicts that under certain conditions liquid-like floc structures may coexist with the usual ordered and disordered phases, there is no experimental observation of this coexistence for electrostatically-stabilized dispersions.

A.3 Colloidal Crystals As Model Many-Body Systems

Since colloidal dispersions permit direct optical examination of their microstructure under suitable conditions, they can be used as model 'many-body systems' to study many-body interactions in materials that cannot be examined directly in a convenient fashion. In particular, formation of order itself offers intrinsic experimental and theoretical advantages, since colloidal crystals can be studied directly using Bragg diffraction and theoretical computations can be made on the basis of known (and directly observed) crystal structures. These potential advantages are slowly becoming practical realities as procedures for producing well-defined structures and instrumental techniques for studying them experimentally are developed. In this Section we have summarized some advances and ongoing research on this aspect.

The following summary focuses only on colloidal crystals. Related examples exist in the case of disordered (i.e., fluid-like) dispersions and will be discussed in the relevant sections of this Chapter and the next.

(i) Crystalline structures are particularly useful as model many-body systems because of their well-defined local structure. However, reliable procedures for growing large, single crystals of known lattice structure are needed for this purpose. One such procedure, for producing single body-centered cubic crystals having specific orientations, has been reported recently by Clark, Hurd and Ackerson (1979). This paper also reports a preliminary study of the quality of the crystal formed. Using laser light scattering, the above authors have examined the presence or the absence of Bragg spots, the spatial distribution of the Bragg diffracted light and the Kossel lines. [A discussion of the meaning and the use of Kossel lines is given in Pieranski (1983).]

(ii) Colloidal crystals melt under shear, and the structural transitions they undergo can be studied using light scattering as illustrated in Figure 6.A.9. Some initial work on shear-induced melting of equilibrium body-centered cubic Wigner lattices was reported originally by Ackerson and Clark (1981). Recently, Ackerson and Clark (1983) have presented an extensive review of these transitions (in both crystals and liquid-like structures) and some theoretical models to explain the observed intermediate structures.

(iii) Colloidal crystals, when confined to a two-dimensional region, offer a unique way of studying many-body interactions in lower dimensions. For instance, the possibility of creation of dislocations and of their observation by optical microscopy by confining colloidal crystals in wedge geometry is an example of such an opportunity [see Pieranski (1983)]. If suitably repulsive boundaries (say, charged glass plates) are used, one can slowly bring the boundaries closer together to form thin layers of 'two-dimensional' crystals or create a wedge geometry to form transitional structures from two to three dimensions, as mentioned above. Pieranski (1983) presents optical photographs of wedge-like crystals and discusses a few additional possibilities, such as creation of thin films and two-dimensional crystals at air/liquid interfaces.

(iv) Recently, model colloidal crystals in binary systems have also been produced artificially [Hachisu and Yoshimura (1980); Yoshimura and Hachisu (1983)]. These binary crystals display ordered lattices of particles of one size imbedded in superstructures formed by the particles of the second size, and are offered as potential macroscopic models of unusual opals formed by silica spheres of two different sizes. Two such binary structures, corresponding to crystalline forms of $NaZn_{13}$ and $CaCu_5$, have been reported by Hachisu and Yoshimura (1980), and are shown in Figures 6.A.10 and 11. Other possibilities are also discussed in the above papers.

The above descriptions give a flavor of the research opportunities that have opened up in recent years on colloidal interactions in general and on many-body interactions in particular. It is clear that research in these areas promises to have impact both on supramolecular systems as well as on their molecular analogues from the fundamental as well as the practical side.

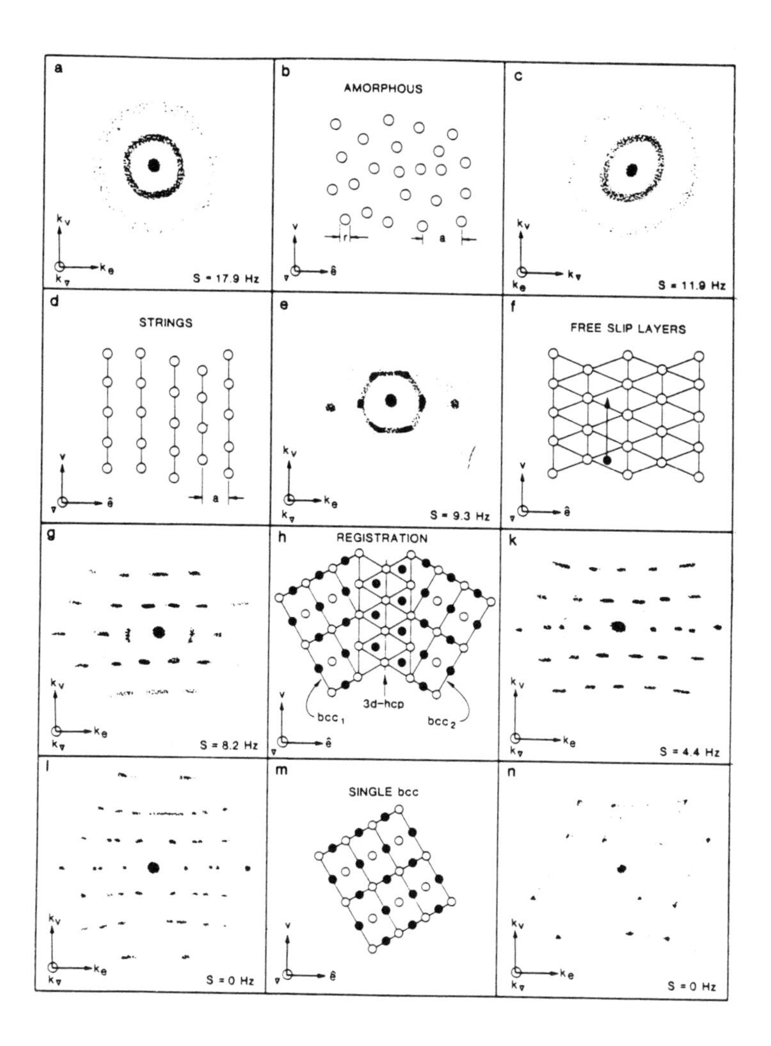

Figure 6.A.9 Scattered intensity distributions produced by a colloidal dispersion [for details, see Ackerson and Clark (1983)].

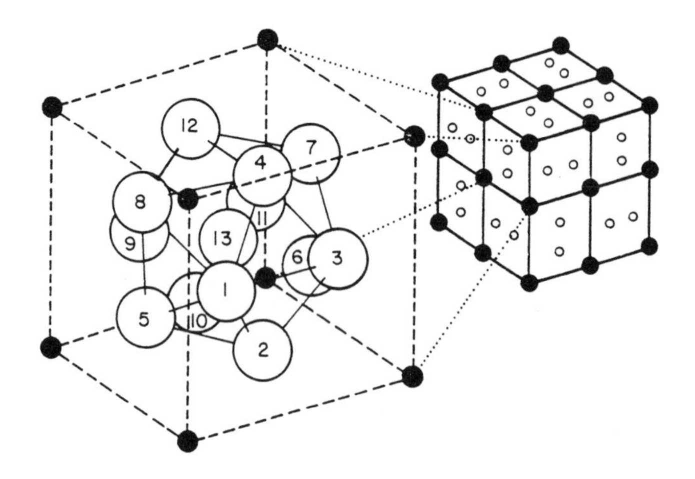

Figure 6.A.10 Binary colloidal crystal with structure similar to that of $NaZn_{13}$ [from Hachisu and Yoshimura (1980)].

The dark circles in this figure denote the positions of the larger of the two types of particles used in forming the crystals (5500 Å and 2700 Å in diameter).

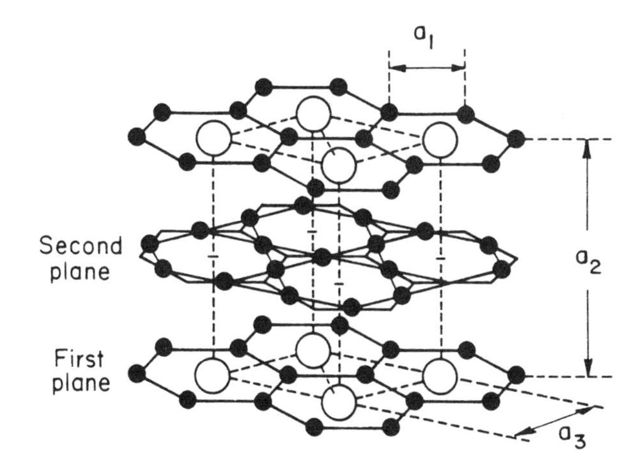

Figure 6.A.11 Binary colloidal crystal similar to $CaCu_5$ [see Hachisu and Yoshimura (1980)].

The open circles correspond to particles 5500 Å in diameter and the dark circles to 3100 Å particles. The inter-planar distance has been exaggerated to reveal details.

B. Static Structure: Colloidal Fluids

It is evident from the previous section that there are numerous problems that require solutions before the crystalline state can be understood adequately and used reliably; nevertheless, the colloidal crystals are generally easier to study than the disordered state because of the relative simplicity of describing the geometric arrangements of the particles in crystals. Thus, the central issues of major concern in the study of interacting, fluid-like dispersions are the theoretical description and the experimental measurement of the local structure. Since much of the physical picture necessary to understand many-body effects in this case has been already set forth in the previous section, our main focus in Section B will be on the theoretical description and use of the local equilibrium structure for interpreting microscale interactions and macroscale properties. Recent developments on the experimental techniques are described separately in Section D of this Chapter, although the relevant experimental observations themselves are integrated with the discussions in this section and in the next section, on dynamic structure.

The most direct piece of information on the equilibrium state (of an isotropic dispersion) that can be obtained theoretically is the radial distribution function, $g(r)$, which is a measure of the local density of the dispersion as a function of distance from the center of an arbitrary particle. The physical significance of the radial distribution function (RDF) is illustrated schematically in Figure 6.B.1. [Alternative (but equivalent) definitions of the RDF are common in the literature; see Watts and McGee (1976).] The RDF is also directly accessible experimentally as structure factor $S(K)$, through suitable radiation scattering techniques (see Figures 6.B.2 and 3; see, also,

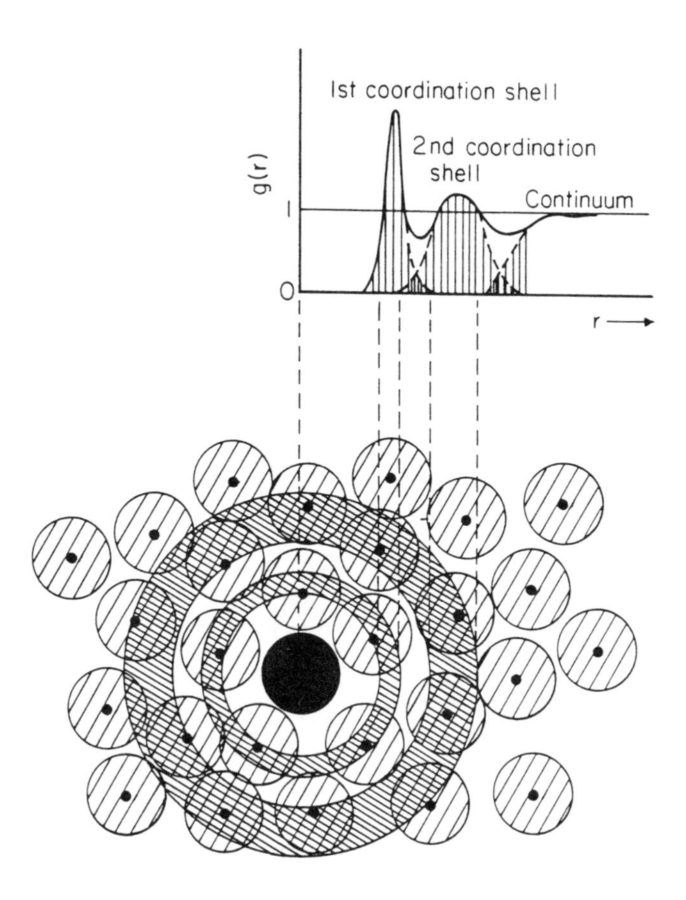

Figure 6.B.1 A schematic representation of the radial distribution function [Ziman (1979)].

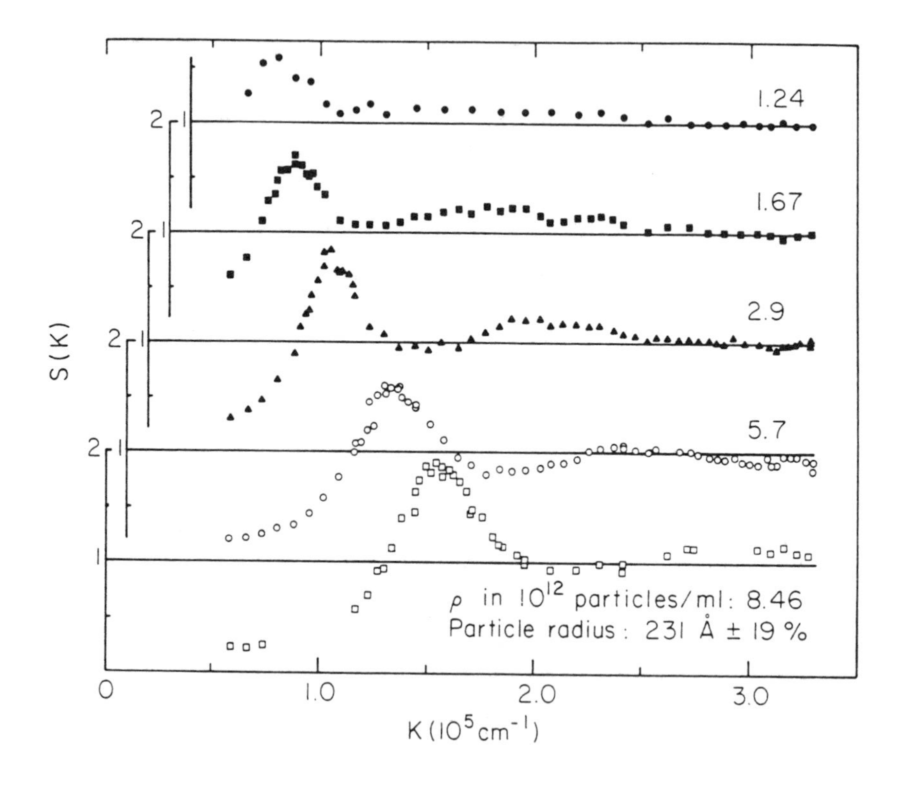

Figure 6.B.2 Development of structure as a function of particle concentration in a charged colloidal fluid, shown in terms of the structure factor (experimental data); data of Brown et al. (1975); see, also, Castillo, Rajagopalan and Hirtzel (1984).

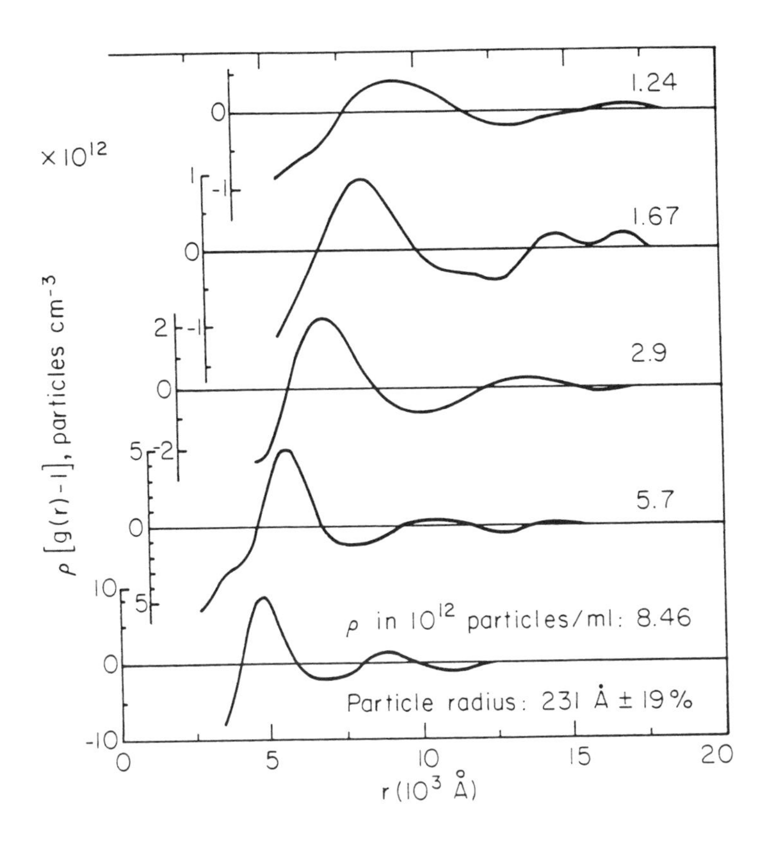

Figure 6.B.3 The radial distribution functions corresponding to the structure factors shown in Figure 6.B.2 [from Brown et al. (1975)].

Section D below). The structure factor is related to g(r) through Fourier transform [see, for example, McQuarrie (1976); the variable K is the magnitude of the scattering vector and is a function of the angle of measurement and the wavelength of radiation used]. The RDF, in combination with the pair potential of interaction, can then be used to obtain the equilibrium thermodynamic properties (such as free energies, osmotic pressures and isothermal compressibilities) of the dispersion. Within this general framework, the major concerns are:

(i) Developing theoretical procedures for formulating radial distribution functions and equilibrium properties in terms of pair potentials of interactions and the relevant physicochemical parameters.

(ii) Developing methods for extracting electrochemical or physical parameters of the dispersions (e.g., intra-micellar structure, aggregation number for the micelles, charge on the particle, etc.) from equilibrium (and dynamic) structure.

(iii) The effects of concentration on the nature of interparticle interactions.

(iv) Effects of polydispersity, shape of the particles and similar factors.

Much of the momentum for the increased research activity in this area comes from recent advances in the theory of atomic and molecular liquids; however, significant differences between molecular fluids and colloidal fluids exist because of the softness and longer range of the interaction potentials in colloidal systems and because of the ionic association/dissociation phenomena that may make the interaction 'dynamic'. In this section, we review the advances in the equilibrium statistical mechanical theories of interacting dispersions, using the above identification of major research directions as a guide. In addition to the above four directions, we also summarize the recent attempts to use fluid-like dispersions as model many-body systems (see also the related review in Section A.3 of this Chapter).

B.1 Liquid-State Theories of Colloidal Dispersions

As mentioned above, the primary objective of the theoretical studies in this area has been to develop procedures for constructing the radial distribution function in terms of the relevant properties of the dispersion or to develop ways of relating the equilibrium properties to these parameters directly. Despite the differences between classical fluids and colloidal fluids mentioned above, the basic formalisms of the theoretical procedures follow directly from liquid-state physics, and excellent references on these are readily available [for example, Barton (1974), Croxton (1974), Barker and Henderson (1976), Hansen and McDonald (1976), Reichl (1980), and the books cited earlier in this section]. Further, semi-technical overviews of the basic physical concepts have been presented recently by Barker and Henderson (1981) and Chandler, Weeks and Andersen (1983). The initial attempts in using the methods of liquid-state physics in the study of dispersions were exploratory in nature, with the goal of examining what could be learned about the equilibrium structures and order/disorder transitions using these techniques. However, in recent years, more attention is being directed to refining these techniques in order to adapt them to the special requirements of colloidal fluids. Roughly, the methods available for constructing radial distribution functions or for calculating properties directly can be classified as either approximate theoretical methods or 'exact' computer 'experiments'. The former are analytical and include integral equation techniques and perturbation theories. The computer experiments, on the other hand, are extensive numerical simulations that supply exact results for testing the theories and fill an important gap in the involved process of developing the approximate analytical procedures into practical and usable tools. The equilibrium structure and properties are generated using Monte Carlo experiments of various types; the Brownian dynamics experiments, whose purpose is to simulate the time-evolution of the microscale structure, will be discussed in Section C of this Chapter.

The following is a brief overview of the progress of these theoretical developments. All through the following discussion, numerous references to methods and approximations named after the originators appear (e.g., Percus-Yevick equation, Born-Green integral equation, etc.); this is frequently unavoidable, as the physics and mathematics literature is replete with eponymy. These usages are standard and are described in the books and papers cited above [see, also, Castillo, Rajagopalan and Hirtzel (1984)]. No attempt will be made to expand on the methods or assumptions involved in any of these cases, unless such additional details are necessary to bring out the physicochemical aspects of the phenomena or to place the technique in proper context with respect to the others.

Integral Equation Methods and Perturbation Theories

One of the earliest attempts to apply statistical mechanical theories to dispersions (of proteins or macromolecules, in this instance) seems to have been made by Kirkwood and Mazur (1952), who use the Born-Green formulation of the integral equation for the radial distribution function. A Yukawa-form of screened interaction potential (with no attraction) is used for the pair potential, and the radial distribution functions are obtained as functions of the particle radius, volume fraction, surface potential and Debye length. Kirkwood and Mazur discuss briefly the possibility of liquid-like local ordering and long-range (crystalline) ordering as the volume fraction approaches and exceeds a critical value. These conclusions are qualitatively correct, but are quantitatively restricted to thick double layers because of the form of the pair potential chosen. More general treatments invariably require numerical solutions, and one such case is provided by Keavey and Richmond (1976), who solve numerically the Ornstein-Zernike integral equation under the Percus-Yevick closure condition for polystyrene latex particles interacting through more general potentials than those assumed by Kirkwood and Mazur. In particular, the interaction potential used includes unretarded van der Waals interaction, a hard-core Born repulsion and an electrostatic potential valid for thin double layers. In addition to radial distribution functions, Keavey and Richmond (1976) also present computed values of osmotic pressures and first-coordination numbers (number of nearest neighbors; see Figure 6.B.1).

The above papers illustrate the applicability of the statistical mechanical theories of liquids to colloids, but the accuracy or the range of applicability cannot be commented upon until exact computations are made using numerical simulations (see below). However, Schaefer (1977) has made a direct comparison between some experimental structure factor data (on polystyrene latex dispersions) and theoretical computations based on Hypernetted Chain (HNC), Percus-Yevick (PY) and Kirkwood-Superposition (KS) approximations. The KS approximation is the same as the one used earlier by Kirkwood and Mazur (1952) as the closure condition for the Born-Green integral equation. (Roughly, the Born-Green-Yvon hierarchy of integral equations for the RDF is exact, but require higher-order correlation functions for completing the equations, i.e., for well-posed equations. The KS approximation is one such closure condition.) Monte Carlo computer experiments on hard and soft (intermolecular) potentials have established that the KS approximation is poor. Nevertheless, the volume fractions for which the KS approximation ceases to converge have been found to be very close to the solid-liquid transitions in the case of hard-sphere fluids, thereby leading to the speculation that the densities at which the KS approximation fails to converge represent the transition points. Schaefer's calculation of melting temperature using the KS approximation is about 30% below the experimental value. Schaefer has also computed structure factors using HNC and PY closure schemes (for the Ornstein-Zernike equation), and the one based on the HNC scheme is closer to the experimental measurements. Because of the previously-established consistency of the HNC scheme with Monte Carlo results on soft coulomb potentials [Hansen (1973); Ng (1974)], Schaefer concludes that the HNC scheme is superior and that the effective charge used in the Yukawa-form of the potential in his calculations can be determined from such comparisons between theory and experiment. Nevertheless, these observations are not conclusive for reasons mentioned later in this review, although an approach that is somewhat similar to Schaefer's, discussed in the following paragraph, has been used successfully by Hayter and Penfold (1981a) to determine charges on colloidal particles.

Perhaps the most noteworthy development in this direction in recent years is the application of the integral equation approach to get an analytical solution for the structure factor by Hayter and Penfold (1981a,b). The application of the Hayter-Penfold solution is described in Section B.2; here we merely note the conditions under which their solution holds. The Hayter-Penfold solution, obtained by solving the Ornstein-Zernike equation in the mean spherical approximation [i.e., another closure relation; see Lebowitz and Percus (1966)], is valid only for macroions interacting through a purely-repulsive potential of the screened Yukawa-form [the type considered by Kirkwood, Mazur and Schaefer previously; the unscreened case can also be solved exactly; see Palmer and Weeks (1973)]. For instance, it is not applicable for the general case of thin double layers (with logarithmic potentials) or attractive interaction potentials; some empirical exceptions to these restrictions and some additional references are given in Section B.2.

Before proceeding to other approximate analytical theories, a few observations that summarize some of the limitations of the above methods are in order.

(i) In addition to the approximations involved in the closure schemes, pairwise additivity of interaction potentials is common to the procedures discussed above. For long-range coulombic interactions, the latter assumption may lead to substantial errors at low ionic strengths.

(ii) Improvements in agreement with experiments are probably not sufficient reasons for choosing one closure scheme over others, unless other, independent reasons exist for the assumptions made regarding, for instance, interaction potentials (e.g., selection of a particular form of repulsion or omission of attractive interactions). Other experimental uncertainties, such as polydispersity, may also lead fortuitously to better agreement with theory.

(iii) Systematic comparisons of exact computer experiments with approximate theories are needed and will reduce some of the difficulties mentioned above.

Some of the computer experiments that have been developed to address the last point will be discussed shortly. Prior to that, another very promising analytical technique, based on perturbation expansions of the properties of the actual colloidal fluid with respect to a suitable reference fluid, deserves attention.

Figure 6.B.4, which presents the structure factors and the corresponding radial distribution functions for a hard-sphere fluid and a soft-sphere, screened-Yukawa fluid at the same volume fraction, shows that the general qualitative features and many of the quantitative aspects of the equilibrium structure of the latter can be reproduced by that of the hard-sphere dispersion under suitable conditions. The hard-sphere structure factor in the figure is computed from the Percus-Yevick equation [Ashcroft and Lekner (1966)], and that of the Yukawa fluid (at a surface potential of 72 mV) follows from the Hayter-Penfold solution. Figure 6.B.4 emphasizes the fact that the hard core of the interaction potential has a strong influence on the local structure, at least under certain conditions. This observation essentially forms the basis of a technique known as perturbation method, in which the thermodynamic properties (or the radial distribution function) of the dispersion are written in terms of a perturbation series expanded about the properties of a suitably chosen reference fluid. The pair potential is usually decomposed into a steep 'reference' part and a longer-ranged and weaker 'perturbation' part (see Figure 6.B.5), and the hypothetical fluid interacting through the reference potential is defined as the reference fluid. The reference fluid is generally approximated by a hard-sphere fluid since the latter is one of the most studied systems in the literature. However, this choice is not always the best; for instance, the long-ranged, slowly decaying coulombic potentials are much too soft to be easily represented by an effective hard-sphere potential. Excellent outlines of perturbation theories and detailed discussions of the finer points can be found in the references cited in the beginning paragraph of Section B.1 and in Barker and Henderson (1967a,b, 1971a,b), Andersen, Chandler and Weeks (1971, 1972) and Henderson and Barker (1970). Application of perturbation theories of both the Barker-Henderson type and the Weeks-Chandler-Andersen type to colloidal fluids has been proven to be very useful [van Megen and Snook (1978); Castillo and

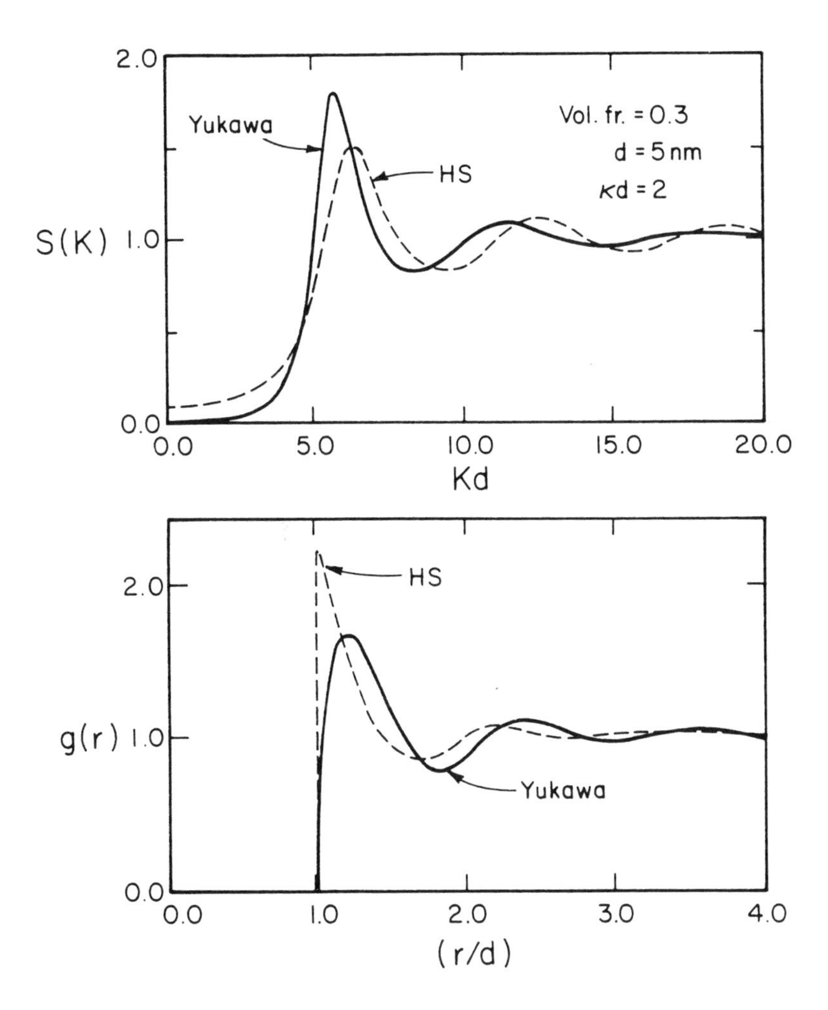

Figure 6.B.4 The computed structure factors and radial distribution functions for a Yukawa fluid and a hard-sphere fluid at the same concentration [see Hayter and Penfold (1981b) and Castillo, Rajagopalan and Hirtzel (1984)].

The inverse screening length is denoted by κ.

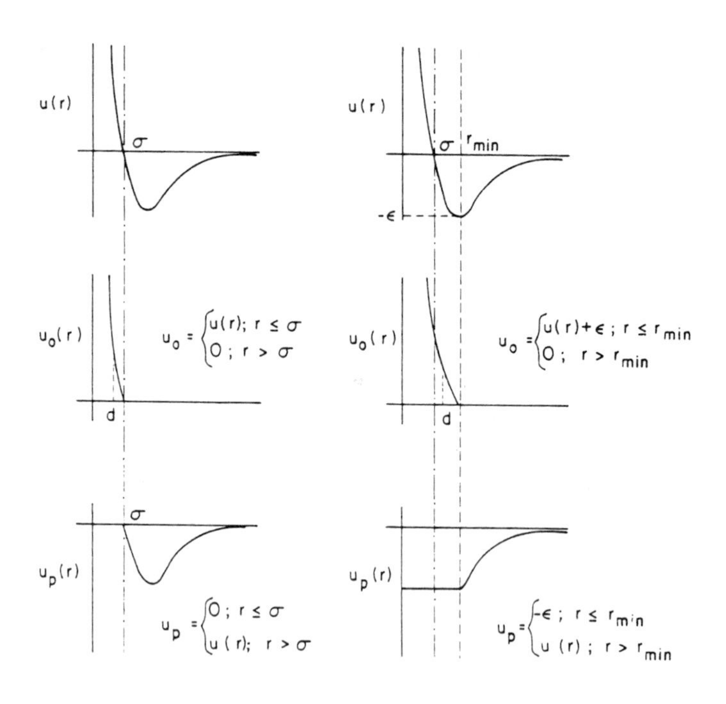

Figure 6.B.5 Two well-known examples of choices available for the selection of reference and perturbation potentials [see Castillo, Rajagopalan and Hirtzel (1984)].

Rajagopalan (1983); Hirtzel and Rajagopalan (1983)]. For instance, osmotic pressures, isothermal compressibilities and radial distribution functions have been computed using perturbation theories and are found to be in good agreement with 'exact' Monte Carlo results when the repulsive part of the potential is steep and the secondary minimum is small; see Figure 6.B.6 [Castillo (1983)]. In addition, the effective hard-sphere diameters computed on the basis of perturbation theories are generally uniformly reliable for estimating the boundaries of order/disorder transitions over the full range of electrolyte concentrations (see Figure 6.A.3 in Section A.1) and are better than ad hoc procedures (such as those of Barnes et al. or Wadati and Toda discussed in Section A.1). Even simpler, first-order perturbation theories of the Barker-Henderson-type offer better estimates of osmotic pressures and 'gelation' conditions than the usual hard-sphere approximations used in application areas such as membrane filtration [see Hirtzel and Rajagopalan (1983)]. More work, however, is needed on the application of perturbation theories to colloids to investigate the effects of softer repulsive potentials and stronger and longer-ranged attractive potentials. The first direct determination of the Helmholtz free energy using non-Boltzmann sampling in computer experiments (i.e., using 'umbrella sampling') has been presented recently by Maleki, Hirtzel and Rajagopalan (1983); such experiments can be used to test the perturbation theories and are described further in the discussion on computer experiments. Finally, a recent work that derives expressions for the osmotic pressure using the random phase approximation (RPA) for the direct correlation function and structure factors [see Hansen and McDonald (1976)] deserves mention. Grimson (1983) uses the Weeks-Chandler-Andersen division of pair potential in the RPA to obtain the long-wavelength limit of the structure

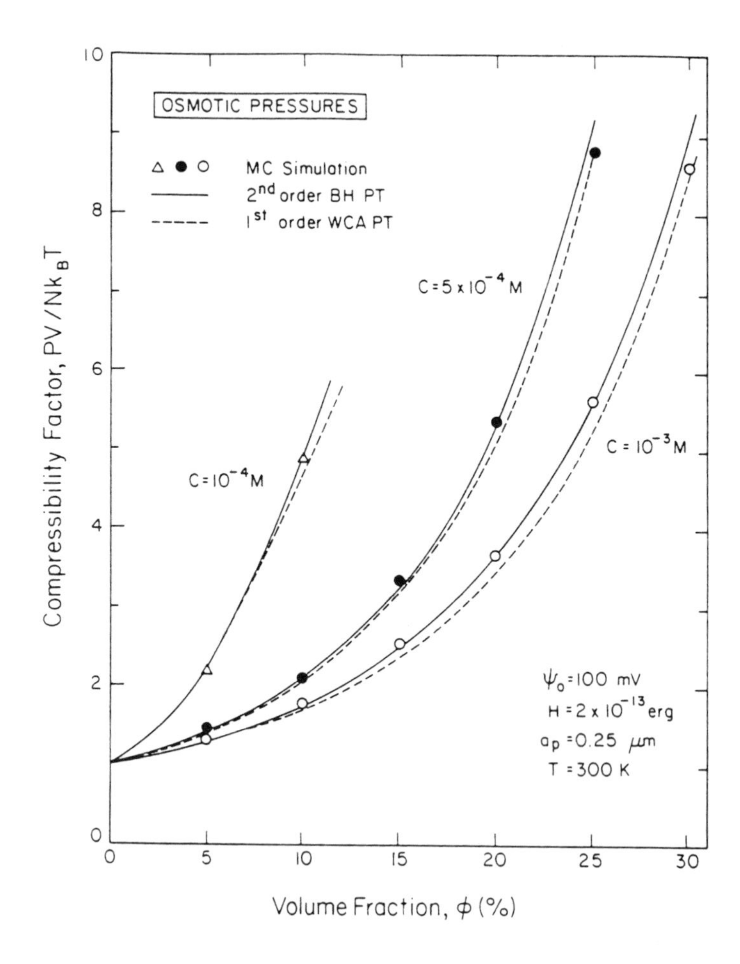

Figure 6.B.6 Comparison of pressures calculated using two types of perturbation theories and computer experiments [see Castillo (1984) and Castillo, Rajagopalan and Hirtzel (1984)].

factor (i.e., for low scattering vectors) and to derive equations for isothermal compressibilities for a Yukawa fluid with a long-range asymptotic form of van der Waals potential. Following the suggestion of Long, Osmond and Vincent (1973) and Cowell and Vincent (1982), Grimson postulates the existence of two-phase regions consisting of a vapor-like, single-particle phase and a liquid-like phase and constructs the corresponding phase diagrams. This leads to a prediction of the existence of three phases (including crystal-like ordering). Although this possibility cannot be ruled out, no experimental observations confirming this have been reported in the literature for charge-stabilized dispersions. Grimson's numerical estimates of osmotic pressure, chemical potential and isothermal compressibility may not necessarily be accurate because of the approximate forms of the potentials used in his theory; however, his estimates of the Hamaker constants, based on his computed isothermal compressibilities for the microemulsions studied experimentally by Vrij et al. (1978), seem to be within the range expected.

Computer Experiments on Colloidal Fluids

As mentioned previously, computer experiments provide 'exact' results [RDF (see Figure 6.B.7) and equilibrium properties] within the restrictions imposed by the usual assumptions (such as pair-additivity of the potentials, which is also made in most of the analytical theories) and hence can be used to test the accuracy of the approximations made implicitly or explicitly in the analytical techniques. Some of the first computer simulations of equilibrium colloidal systems were presented by van Megen and Snook (1978), who compared osmotic pressures and free energies obtained from perturbation theories and cell models with the exact results based on Metropolis Monte Carlo experiments (see also the papers of van Megen and Snook cited in Section A of this

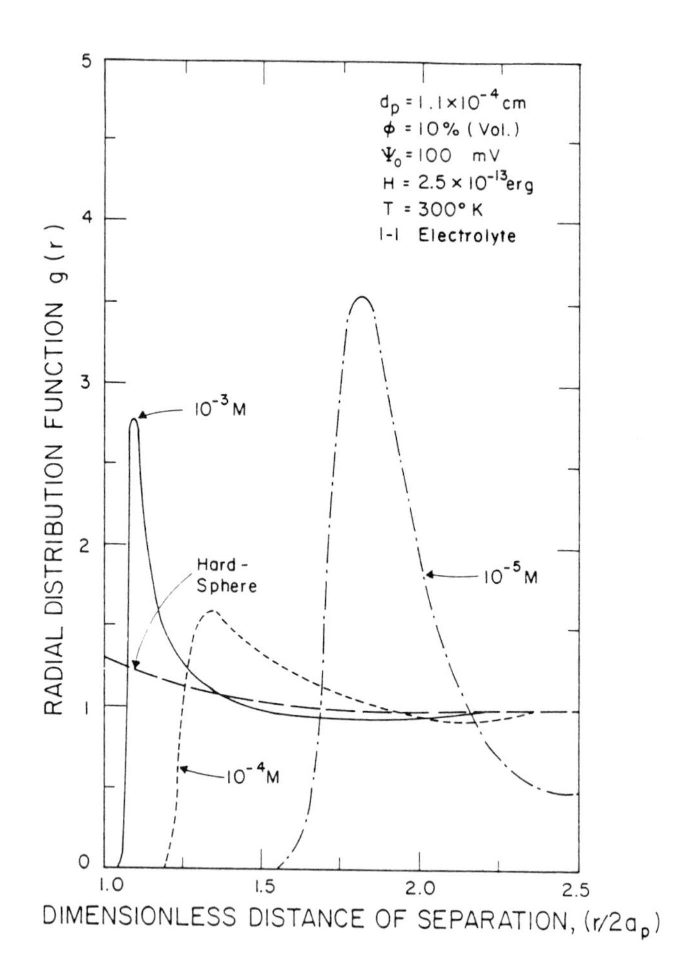

Figure 6.B.7 Radial distribution functions computed using Monte Carlo simulations [from Castillo, Rajagopalan and Hirtzel (1984)].

The parameters of the dispersions are shown in the Figure.

Chapter). The results of the simulations generally support the predictions of the approximate theories; however, extreme conditions (in volume fractions, electrolyte strength, softness or range of the pair potentials) have not been tested thoroughly yet, and recently, Castillo (1984) has undertaken a broader test of perturbation theories using computer experiments [see also Castillo (1982) in Section A]. The Metropolis Monte Carlo method [see Metropolis et al. (1953)] is not reliable near two-phase regions because of the small number of particles to which one is invariably restricted in the simulations. In addition, it cannot be used to obtain the free energies directly because of the sampling procedure used [known as 'importance sampling', which restricts the configurational energies sampled to the most probable, low energy intervals, whereas the high energy interval is important for determining the partition functions and free energies; see Valleau and Torrie (1977)]. The first step in obtaining free energies directly has been taken by Maleki, Hirtzel and Rajagopalan (1983), who employ a weighted-Boltzmann sampling known as 'umbrella sampling' to compute _excess_ free energies (over that of a reference dispersion). These computations are in good agreement with the results of first-order Barker-Henderson theory. Maleki, Hirtzel and Rajagopalan (1983) also demonstrate that under certain conditions one can use Metropolis importance sampling to determine excess free energies directly by choosing the reference system suitably.

The above studies are only the first step in the overall effort to build a comprehensive understanding of interacting dispersions. Some additional work needed in these directions is listed at the end of this Chapter. Finally, it is also important to note that the discussions in this section and in what follows are relevant in the case of _steric_ systems as well.

Steric systems display a richer variety of phenomena because of the complex role played by the polymeric macromolecules that are used to impart stability or instability (see Section B of Chapter 4 for additional information). A brief discussion of this point and additional references are given in Castillo, Rajagopalan and Hirtzel (1984).

B.2 Use of Theories of Many-Body Interactions For Determining Parameters of the Dispersions

As mentioned in the introduction to Section B, considerable progress has been made on developing methods for extracting electrochemical and physical parameters of the dispersions from a combination of experimental data and theories of many-body interactions. An excellent example of the need for includin~ long-range interactions and interparticle correlations in the analysis of scattering data is provided by Hayter and Penfold (1981a,b) in their study of intra-micellar structures of sodium dodecyl sulfate (SDS) micelles. These papers demonstrate, using small-angle neutron scattering data and computations of structure factors in terms of interaction potentials, that the charge and the aggregation number of SDS micelles can be determined reliably if the many-body effects are included in the analysis. The Hayter-Penfold analysis thus illustrates, simultaneously, both the need for studying interaction effects as well as some of the practical information that can emerge from such an analysis. The Hayter-Penfold scheme, which applies to purely-repulsive potentials, can be modified to study the influence of attraction if the repulsive core of the potential is steep and if the attractive tail can be approximated by an effective-Yukawa potential. Hayter and Zulauf (1982) have taken advantage of this to study the structure of solutions of n-octyl pentaoxyethylene glycol monoether, where the above two conditions are met reasonably adequately. Some additional information on these is also available in Hansen and Hayter (1982) and Hayter (1983). Although the potential used by Hayter and Penfold (1981b) in deriving the analytical result for the structure factor is appropriate only for thick double layers, Bendedouch and Chen (1983) have shown that this restriction can be circumvented empirically by equating the actual structure factor to that of a hard-sphere dispersion with the same second virial coefficient. The values of S(K) calculated in this manner seem nearly identical to the Hayter-Penfold results for low micellar concentrations. One reason for this equivalence is that the Yukawa potential (used by Hayter and Penfold) and the actual logarithmic potential (of Bendedouch and Chen) are very nearly the same for separations larger than the equivalent hard-sphere diameter. Additional details on the experimental systems used by Chen and co-workers are given in Bendedouch, Chen and Koehler (1983).

Extraction of effective pair potentials from observed structure factors has also been attempted with limited success. Methods based on integral equation theories, such as Percus-Yevick theory and Hypernetted Chain (HNC) theory, have been tried recently by Nieuwenhuis and Vrij (1979) in the case of polymethylmethacrylate latex dispersions in benzene; but these methods are unreliable at large concentrations or for strongly-interacting dispersions. In fact, this deficiency has been observed from computer experiments on atomic and molecular fluids [Ailawadi (1980)]. In view of this weakness, the observation of Nieuwenhuis and Vrij that the extracted pair potential was much too similar to the potential of mean force is not surprising.

B.3 Interaction Potentials in Concentrated Colloidal Dispersions

The nature of electrical double layer interactions in strongly-interacting dispersions has been discussed in considerable length in Section A.1 in the context of ordering in supramolecular systems. In the present section, we draw attention to a few additional investigations on this topic, which are relevant to both ordered and disordered systems. The liquid-state theories of colloidal fluids discussed in the previous sections require the form of the interaction potential as input. For this, one usually uses the conventional DLVO potential (see Chapters 3 and 4), which asssumes the surface of the colloidal particle to be in osmotic equilibrium with an infinite electrolyte reservoir at known or calculable concentration. On the other hand, in practice, treatment with ion-exchange resins and subsequent addition of salts leave the properties of the reservoir undetermined, although in principle equilibrium dialysis can be used to determine these. To circumvent this, Beresford-Smith and Chan (1982, 1983) treat the dispersion as a highly asymmetric electrolyte (of known total composition; information on the reservoir is not needed). Some studies along this line have been cited in Section A.1, and others [see Stephen (1971) and Harris (1975)], which examine the dispersions in the linear Debye-Hückel limit, are also available in the literature. Since the linear Debye-Hückel limit is inappropriate for highly-charged particles, Beresford-Smith and Chan (1982, 1983) use an asymptotic analysis of the Ornstein-Zernike equation for a multicomponent system (in the first paper) and an alternative approach that regards the colloidal particles as 'solutes' in a 'solvent' of small ions (in the second paper). A cell model, in which each colloidal particle is confined to an electrically neutral spherical cell with an average number of counterions and added electrolyte ions, is also available in the literature [Beunen and White (1981)]. The nonlinear Poisson-Boltzmann equation is solved within the cell to obtain the effective potential. The effective pair potentials obtained by each of these approaches have been used to compute the structure factors, by Beresford-Smith and Chan (1983), who then compare these with the experimental data of Brown et al. (1975). The structure factor based on the cell model differs substantially from that based on the Beresford-Smith/Chan approach. The latter is in reasonable agreement with the experimental data of Brown et al. (1975). It however appears that the equilibrium structure factor alone is insufficient to discriminate between the various approaches; dynamic intermediate scattering functions and diffusion data might improve the situation.

B.4 Effect of Polydispersity

Polydispersity, of size as well as of charge and of potential at the surface of the particles, is known to be important in determining the properties and the structure of the dispersion, but very little theoretical or experimental work has been done to study it systematically. Thermodynamics of polydisperse hard-sphere and hard-disc systems has been studied extensively, and a good review and some recent results are available in Münster (1974) and Salacuse and Stell (1982), respectively. Dickinson (1983) has also reviewed some additional investigations that may be useful in the study of colloidal fluids. Hard spheres of different sizes pack more efficiently than a monodisperse system, and consequently, at constant number density, the osmotic pressure of a polydispersed suspension is lower than that of monodispersed particles at the same volume fraction [Dickinson (1978)], although the difference is rather small for low volume fractions (below about 0.2). In the case of interacting dispersions, the difference is even smaller and negligible in the disordered state, but becomes large and positive in the amorphous, solid-like state [Dickinson (1979); Dickinson, Parker and Lal (1981)].

Repulsive interaction tends to keep the particles apart so that the correlation between position and size (and, in fact, shape) is minimized in the disordered state. Further, the net effect of polydispersity on the order/disorder transition is towards the disruption of long-range order. It is also clear from the work of Dickinson, Parker and Lal (1981) that the radial distribution function alone is not a good measure of the structure (i.e., the 'solid phase' has a glass-like, amorphous structure rather than a crystalline structure). Finally, it is worth noting that the 'permeable' sphere model of Blum and Stell (1979) and Salacuse and Stell (1982), in which the particles are allowed to overlap with a finite probability, may provide a useful basis for studying microemulsions that can coalesce briefly and then separate. Blum and Stell (1979) and, independently, Vrij (1978a,b) have also derived the structure factor for polydispersed hard-sphere fluids using the Percus-Yevick approximation. The derivation of Vrij (1978a,b) is restricted to the compressibility limit (i.e., low-K limit) of the structure factor, but its finite-angle extension and intensity calculations at finite concentrations have been presented more recently [Vrij (1979); van Beurten and Vrij (1981)]. Extensions of these to interacting colloidal fluids are needed; but since repulsive interaction tends to diminish the effect of polydispersity, the above theories provide a useful upper bound.

B.5 Colloidal Fluids as Model Many-body Systems

In Section A.3 the interesting possibility of using colloidal crystals as model many-body systems to study many-body interactions in supramolecular (and molecular) systems was discussed. This possibility is not restricted to crystal-like structures. In particular, the report of Clark and Ackerson (1980) on an experimental study of spatial structures of spontaneous thermal fluctuations in a colloidal fluid opens up interesting avenues in this regard. This paper describes observations of shear-flow-induced distortions of the static structure factor in strongly-interacting colloidal dispersions maintained in a steady non-equilibrium state. Such observations in molecular systems are impossible in practice because of the extreme experimental conditions required to maintain forced deviations from the equilibrium state. As mentioned in Section A.3, Ackerson and Clark (1983) present some specific examples of the above-mentioned application (e.g., shear-induced melting of crystalline materials and re-entrant transition to translational order at higher shear rates). The examination of these possibilities has just begun, and research on this aspect offers promising new directions.

C. Dynamic Structure of Interacting Dispersions

The term 'dynamic structure' is rather broad and can be defined to include a variety of nonequilibrium phenomena and properties. However, we shall use it in a very restricted sense because of the limitations imposed by the scope and the focus of this Chapter. Specifically, this section will be concerned primarily with the evolution of the local geometric structure of the dispersion from one equilibrium configuration to another. In this sense, the subject matter of this section is, to a large extent, a direct extension of the material discussed in the previous section, on static structure.

Some of the more general facets of dynamic structure, such as rheological properties and phenomena and diffusion coefficients, are discussed in Chapter 7, under Transport Properties. The material reviewed in the present section under dynamic structure has direct relevance to the theoretical and experimental concepts needed for interpreting some of these transport phenomena and properties (particularly, diffusion coefficients). The research activities in this area (especially those that focus on the role of colloidal interactions) are very recent and are concerned primarily with a few fundamental aspects, among which are the determination of time-dependent correlation functions for the positions of the particles, their relation to the scattering functions that represent them in experimental (radiation scattering) measurements, and the influence of direct colloidal forces and the indirect hydrodynamic interactions on the evolution of the configurations. We review the current activities on each of these aspects here. References to some experimental techniques that measure the dynamic correlation functions will be made where necessary. Additional details on the general experimental techniques are given in the following section.

The equilibrium radial distribution function discussed in Section B is in fact a special case of more general distribution or correlation functions that are used to represent the temporal evolution of the spatial distribution of the particles. The generalization of the RDF needed for describing the dynamic structure is known as the Van Hove correlation function [see Barton (1974); Boon and Yip (1980)] and is usually divided into two parts, namely, the self and the distinct correlation functions. The Van Hove self correlation function describes the probability of finding a particle at any position r at time t, given that it was at r=0 at t=0. It represents the evolution of the singlet distribution and, thus, describes the diffusion of the particle away from its initial position. In non-interacting systems this is a Gaussian distribution that flattens with time. The distinct correlation function represents the probability of finding a particle at any position r at time t given that another particle occupied the origin (r=0) at t=0. This time-dependent analogue of the pair-distribution function (i.e., RDF) reduces to the RDF at t=0. A typical example of the Van Hove distinct space-time correlation function, obtained from a computer experiment (discussed later), is shown in Figure 6.C.1. The intermediate scattering functions measured in dynamic radiation scattering experiments (say, by quasi-elastic light scattering; see Section D) are spatial Fourier transforms of the Van Hove functions and, thus, the theoretical description of the Van Hove functions occupies an important position in the study of the dynamics of colloidal fluids.

The research activities on these topics may be classified, roughly, under the following categories.

1. Development of computer experiments to generate 'exact' data on the time-evolution of configurations and distribution functions.

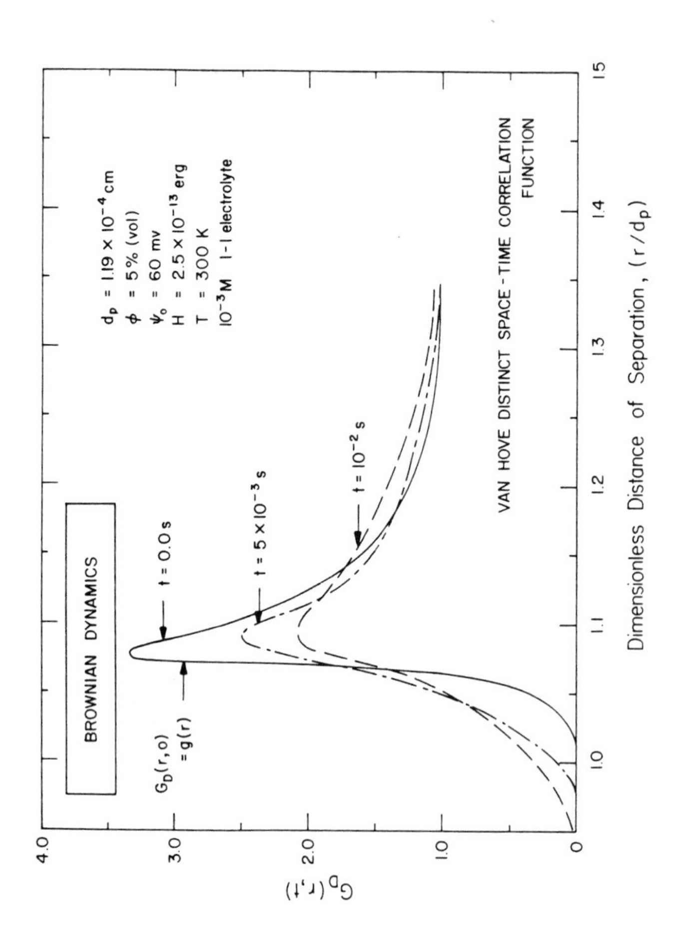

Figure 6.C.1 Van Hove distinct space-time correlation functions computed using Brownian dynamics experiments [see Venkatesan (1984)].

2. Theoretical examination of intermediate scattering functions to obtain their relation to 'free'-diffusion coefficients, self-diffusion coefficients and collective diffusion coefficients.

3. Analysis of effects of polydispersity on scattering and development of methods to obtain size distribution from measured scattering functions.

In addition, as discussed in other sections of this Chapter, some exploratory (theoretical as well as experimental) studies on the dynamic aspects of structural changes under externally-imposed disturbances (such as imposed shear) are also underway. These are reviewed in the following pages.

C.1 Description of Evolution of Local Structure

The Monte Carlo techniques discussed in Section B are designed to generate equilibrium configurations randomly and cannot provide information on the evolution of one configuration from another. The generation of time-dependent configurations requires the solution of the actual equations of motion for the particles. This task, which is the objective of well-known molecular dynamics algorithms used in liquid-state physics [see Alder and Wainwright (1957)], can be accomplished, at least in theory, by what is known as Brownian dynamics method [Ermak and McCammon (1978)] in the case of colloidal particles suspended in a continuum. The Brownian dynamics algorithm of Ermak and McCammon extends the deterministic equations of motion used in molecular dynamics method to include the random Brownian motion of the particles [see also the earlier papers of Ermak and Yeh (1974) and Ermak (1975a,b)]. The resulting system is a set of coupled Langevin equations, whose solutions describe evolution of the velocities and the configurations [see Deutch and Oppenheim (1971)]. If the time steps used in the simulation are much larger than the relaxation time for the Brownian motion of the particles, the equations can be simplified further and written in terms of first-order differential equations for the positions of the particles. Ermak and McCammon (1978) use Langevin equations in the configurations space, but their procedure is consistent with the alternative formalism based on Fokker-Planck equation, whose solution yields the phase-space probability distribution function directly [Murphy and Aguirre (1972)].

One of the major difficulties in the simulation of Brownian dynamics is the incorporation of indirect hydrodynamic interactions among the particles in the equation of motion. This manifests itself in two ways: the first is the problem of theoretically specifying the complex many-body hydrodynamics. Even the simplest case of two-body interactions has been described sufficiently adequately only recently [see Batchelor (1976) and the discussions in Section C.2 of Chapter 4]. The second is the relatively simpler problem of efficiently incorporating the available expressions in the computational procedure. The latter increases the computation time, but does not pose any conceptual difficulties. In the absence of hydrodynamic interactions, Brownian dynamics simulation is relatively simple (albeit time-consuming) and has been presented by Gaylor et al. (1979, 1980), Tutt (1982) and Venkatesan (1984). Results of such simulations can be presented in terms of

a. Van Hove correlation functions,

b. intermediate scattering functions, and

c. diffusion coefficients, computed from mean-square displacements (discussed in Chapter 7).

The first two are equivalent. An example of the computed Van Hove distinct correlation function has been shown in Figure 6.C.1. The self correlation function, which is Gaussian for non-interacting particles, deviates strongly from the free-diffusion case when either the particle concentrations or the colloidal interactions become significant. The Gaussian form is the optimum estimate if only the first two moments (the mean and the standard deviation of the displacements) are known. However, the general case of the self correlation function (for interacting particles) can be expanded in terms of an infinite series (in Hermite polynomials) with the Gaussian form as the reference [Rahman (1964); Nijboer and Rahman (1966)]; this has been attempted by Gaylor, Snook and van Megen (1981), who find marginal improvements through such a series expansion for intermediate times (of the order of a few milliseconds). The Gaussian form becomes exact in the limits $t \to 0$ and $t \to \infty$. One of the traditional goals of the theories of dynamic properties of molecular fluids has been to express the distinct correlation function in terms of the self correlation function. One possible approximation for this relation is due to Vineyard (1958) who equates the distinct correlation function at any time t to the convolution of the radial distribution function and the self correlation function (at the same instant t). Since this has been found inadequate, a 'delayed Vineyard' approximation has been suggested by Nijboer and Rahman (1966); this requires that the self correlation function be taken at time $t' < t$, where t' is determined empirically from exact results for the correlation functions. This also has been tried for a particular case of interacting dispersions by Gaylor, Snook and van Megen (1981), who find that the extensions of these concepts from liquid-state physics are useful for studying colloidal systems. No special practical benefits follow from these results, however, since the delay time t' cannot be determined a priori. In addition, no theoretical justification for these approximations is available currently.

The inclusion of hydrodynamic interactions in the above computations is a formidable task that has not yet been investigated satisfactorily. Usually n-body interactions for n > 2 are assumed to be given approximately by the sum of pair interactions -- an assumption that cannot be justified on physical grounds because of the long-range nature of the hydrodynamic interactions. More specifically, in concentrated dispersions, pair interactions are 'screened' strongly by the intervening particles and the nearest neighbors exert a much stronger influence than the particles farther away. In the absence of rigorous methods for including many-body hydrodynamic interactions, various approximations are being tried currently although the accuracy of these is open to speculation. The approximations that have been tried are the following:

1. Ermak and McCammon (1978) suggest using pair additivity with either the Oseen tensor [see Yamakawa (1971)] or the Rotne-Prager tensor [Rotne and Prager (1969)] for the mobility. The use of these mobility tensors simplifies the Langevin equation since a term containing the gradient of the diffusion coefficients becomes identically zero. This, however, implies that the initial diffusion coefficients are independent of the local configuration, which is not correct (see Chapter 7, Section A). Ermak and McCammon have nevertheless used this approximation to study dimers and trimers.

2. More accurate mobility expressions for a pair of interacting particles have been derived by Batchelor (1976) and can be used in place of the above. Batchelor's results are available numerically and reduce to the Oseen approximation at large separations. The results of Batchelor have been used recently by Venkatesan, Hirtzel and Rajagopalan (1983), Bacon, Dickinson and Parker (1983a,b) and Venkatesan (1984). The first of these papers is on computing initial diffusion coefficients using equilibrium mobility averaging technique and will be discussed in Chapter 7. The others are Brownian dynamics simulations that focus on stability of dimers and trimers [Bacon, Dickinson and Parker (1983a)], stability of concentrated dispersions [Bacon, Dickinson and Parker (1983b)] and correlation functions [Venkatesan (1984)]. The last two also use a cut-off distance for computing hydrodynamic interactions in recognition of the screening effect mentioned above.

3. Conceptually the above two are about the same, with the exception of the fact that Batchelor's results for the mobility tensor offer an improvement over those of Oseen and Rotne and Prager. However, recently Mazur and van Saarloos (1982) have derived a cluster expansion for the mobility tensor that includes interactions via third

and fourth spheres. Their expressions includes terms up to order r^{-7} in the interparticle spacing r and reduces to the earlier derivation of Felderhof (1977), based on the method of reflections, when only pair interactions are considered. These expressions have been used to compute diffusion coefficients [Felderhoff (1978); van Megen, Snook and Pusey (1983); Snook, van Megen and Tough (1983)], which are discussed further in Chapter 7, Section A. The currently available Brownian dynamics results do not use the above mobility tensor.

4. An alternative approach, which uses experimental data to derive an effective two-particle mobility tensor, has been attempted recently by Snook, van Megen and Tough (1983). Since hydrodynamic interactions extend over large enough distances to cause computational problems in computer experiments, this reverse approach to estimating effective mobility tensors from experimentally observed self and collective diffusion coefficients offers a semi-empirical method whose results for the effective mobility can then be used in Brownian dynamics experiments [see, also, Pusey and van Megen (1983)]. In this sense, this method is analogous to extracting effective pair potentials for colloidal interactions from structural data (see Sections B.2 and B.3 above).

The above summary outlines the methods that are available currently for incorporating many-body hydrodynamic interactions (in a necessarily approximate manner) in the study of stability and diffusional behavior of concentrated systems. The current status in the application of these concepts may be summarized as follows.

(i) Brownian dynamics simulations that provide space-time correlation functions have been demonstrated to be possible in principle. These can also be used to generate intermediate scattering functions theoretically.

(ii) The relation between the self correlation function and the distinct correlation function in the case of colloidal dispersions is very similar to that observed in the case of classical fluids.

(iii) The major difficulty that must be overcome is the inclusion of indirect, hydrodynamic interactions in the computations. Pair additivity of hydrodynamic interactions is unlikely to be of acceptable accuracy except at high dilution; this limitation is severe in the case of collective diffusion and sedimentation. A few theoretical and semi-empirical expressions for effective pair interaction are available, but have not been explored sufficiently.

(iv) The computational task is formidable, especially with the inclusion of hydrodynamic interactions. The long range of these interactions imposes severe restrictions on the periodic boundary conditions and box size used in the simulations, although the screening of the interactions by the intervening particles may ease these restrictions.

(v) It is perhaps possible to study the range of many-body hydrodynamic interactions and to examine the available expressions for the mobility by computing diffusion coefficients and comparing these with experimental results (see, also, Chapter 7, Section A).

C.2 Theoretical Examination of Scattering Functions

As mentioned earlier, the space-time correlation functions are related to the intermediate scattering function F(K,t) (measured in dynamic radiation scattering experiments) through spatial Fourier transforms. Quasi-elastic light scattering (QELS) does not measure the self and the distinct correlation functions separately, whereas neutron scattering can be used to observe the self correlation function (through 'incoherent' scattering) separately. As described in Section D below, QELS can provide, essentially, the incoherent scattering function at large scattering vectors (where $S(K) \sim 1$) and the coherent scattering function (i.e., collective diffusion) at low scattering vectors [see Pusey (1978, 1980)]. In view of this, the intermediate scattering functions have been studied directly to see how the diffusion coefficients can be extracted from these experimentally. We describe below the efforts in this direction. These studies are also relevant to the discussion presented in Section A of Chapter 7.

(i) Short-time expansions for the intermediate scattering function have been derived by Ackerson (1976) from the Generalized Smoluchowski Equation and by Pusey (1975) from the Langevin equation [see, also, Pusey and Tough (1982)]. From the time-derivative of this expansion, one can show that the collective diffusion coefficient is given by the ratio of free-diffusion coefficient to the structure factor. (This is for hydrodynamically non-interacting dispersions.) This result is analogous to the similar observation made by de Gennes (1959) for molecular systems.

(ii) The colloidal interactions, which enter through the structure factor in this result, were originally thought to enter extraneously through the normalization of F(K,t) by S(K) in the derivation [see Pusey (1979, 1980)]. This interpretation is, however, incorrect since the change in osmotic compressibility due to colloidal interactions must have a direct effect on the collective diffusion coefficient. [This also follows from the generalized Stokes-Einstein relation; see Boon and Yip (1980).] This has been demonstrated by an alternative derivation of the same result by Ackerson, Pusey and Tough (1982) recently.

(iii) Pusey and Tough (1982) have extended the above derivation by including hydrodynamic interactions. They also demonstrate that for monodispersed systems the short-time and the long-time limits of the collective diffusion coefficient are identical. The self-diffusion coefficient, however, is reduced further at large times because of direct colloidal interactions. The implication of these observations will be discussed in Chapter 7, Section A.

(iv) Experimental scattering data on self-diffusion coefficients and intermediate scattering functions suggest that they probably can be described in terms of the Brownian motion of a typical particle trapped inside a 'cage' formed by its nearest neighbors [see Pusey (1978); Boon and Yip (1980)]. Such cage structures have been employed successfully in the study of simple liquids (such as liquid argon); in these latter cases, harmonic approximations of the potential distribution inside the cages have been shown to predict the self and the distinct Van Hove correlation functions reasonably well [see Tsang (1979); Tsang and Jenkins (1980)]. This has motivated a similar approach to colloidal fluids [Tsang and Tang (1982); Rajagopalan (1982)]. The advantage of this approach is that closed-form solution to the single-particle Brownian motion equation can be obtained and then averaged over all initial positions. The result can then be equated to a short-time expansion of F(K,t) to get the effective scattering function. Values of mean-square displacements computed in this manner have been compared with the 'exact' results from a Brownian dynamics experiment in Figure 6.C.2 to demonstrate what is possible through this approach [Rajagopalan (1982)]. Comparisons with experimental data of Pusey (1978) are shown in Figure 6.C.3 [see, also, Tsang and Tang (1982)]. This approach, however, is not accurate at large times (since the harmonic approximation becomes inaccurate) and is suitable only for short times and for strongly-interacting dispersions.

C.3 Effect of Polydispersity

The effect of polydispersity on the static structure and the thermodynamic properties has been discussed in Section B.4. The literature on this is more

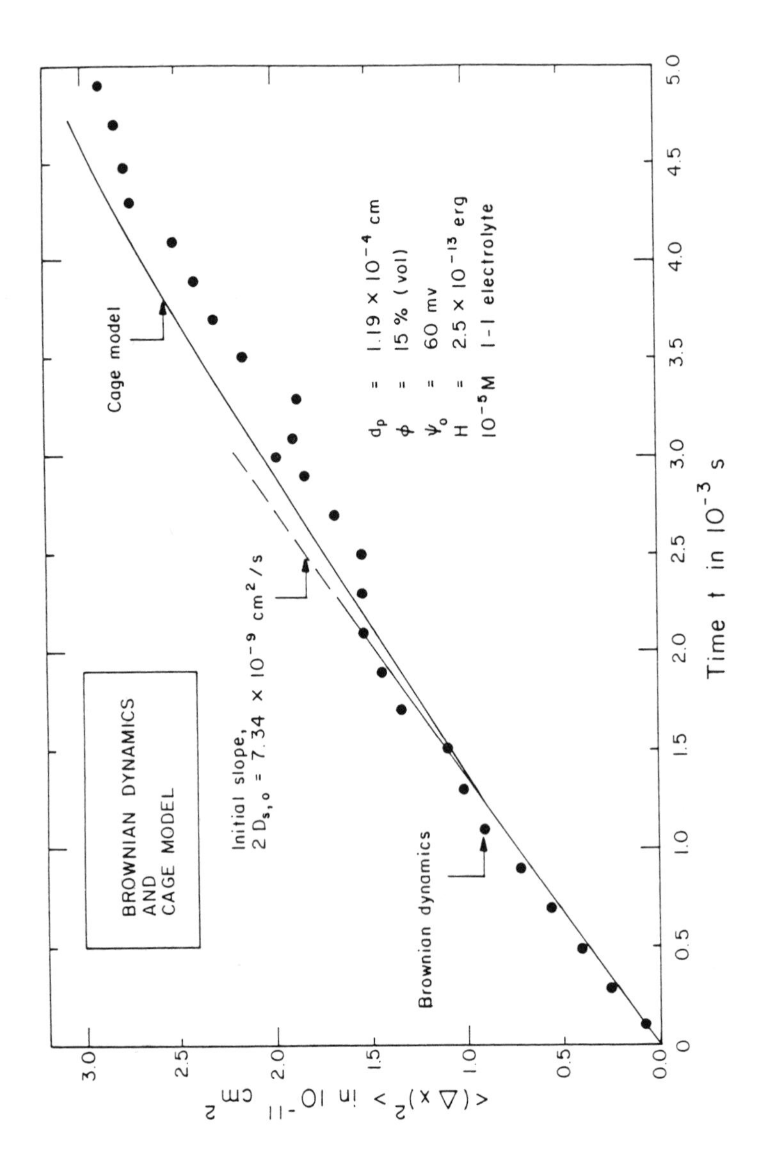

Figure 6.C.2 Comparison of mean-square displacements computed using a cage model and exact Brownian dynamics experiments [from Rajagopalan (1982)].

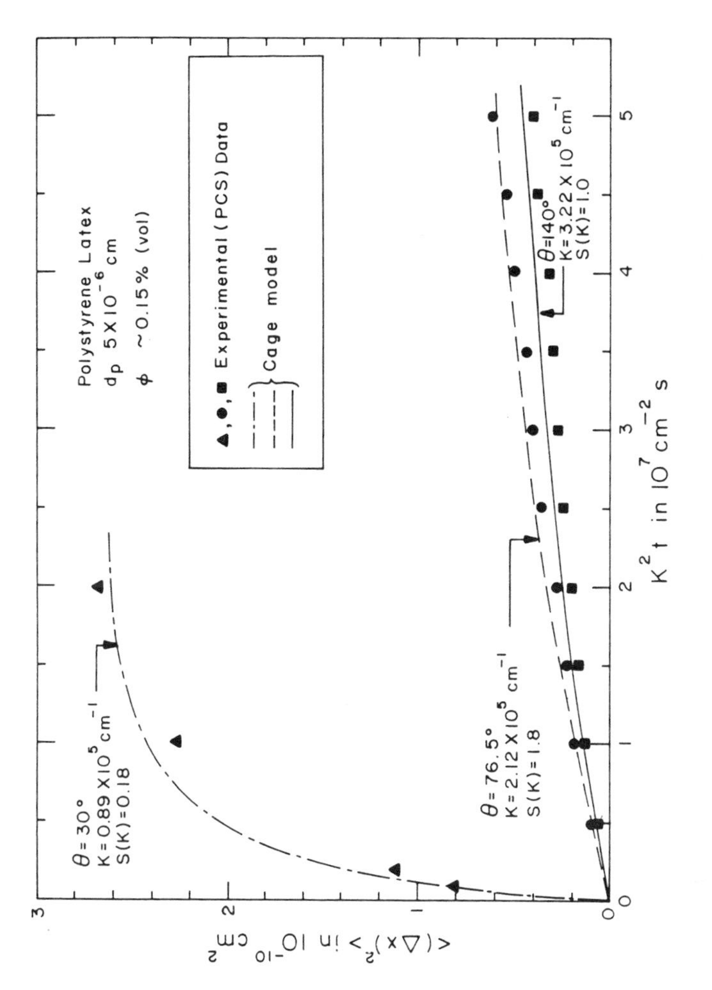

Figure 6.C.3 Comparison of experimental values of mean-square displacement and those based on the cage model [from Rajagopalan (1982)].

extensive than on the dynamic aspects of dispersions. Brownian dynamics simulations on polydispersed suspensions are more difficult because of the poorly-understood hydrodynamic interaction effects; however, because of the obvious practical interest, some results are available on the effect of polydispersity on the intermediate scattering function and particularly on the effective diffusion coefficients. The expression for the effective diffusion coefficient for non-interacting particles is given by Brown et al. (1975) in terms of the appropriate moments of the size distribution. For non-interacting particles, the existence of a small amount of polydispersity does not affect the scattering measurements significantly. In contrast, in the case of interacting particles, the collective diffusion can be completely masked by self-diffusion (i.e., coherent scattering is masked by incoherent scattering) if the particles are polydisperse. This effect of polydispersity was first pointed out by Weissman (1980), who explained why the experimental intermediate scattering function of Pusey (1978) showed slow decay (corresponding to self-diffusion) even at low scattering vectors (where one normally observes faster, collective diffusion). Weissman's observation also explains the slow relaxation of the refractive index fluctuations in dilute, ordered dispersions [see, Weissman, Pan and Ware (1978); Dickinson (1983)]. The self-diffusion coefficient, however, is not affected significantly by small amounts of polydispersity [see also Phillies (1977)]. Pusey (1980) presents additional details on Weissman's approximate equations for the structure factor and intermediate scattering function and notes that the simplified equations for 'paucidispersed' suspensions (i.e., very narrow size distribution around the mean) can be taken advantage of in certain cases. That is, the 'noise' due to polydispersity can be turned into an advantage in the case of dispersions for which high-scattering-vector measurements (i.e., self-diffusion measurements) are physically inaccessible because of low interparticle separations. Such situations occur in some concentrated dispersions of proteins and microemulsions, in which all of the accessible scattering vectors correspond to only the compressibility limit of S(K) and F(K,t). Deliberate 'seeding' of the suspension with some large particles may be used in these cases to measure self-diffusion coefficients. Alternatively, if the measured low-K diffusion coefficient is known to be the self-diffusion coefficient, the equations of Pusey (1980) and Weissman (1980) can be used to obtain some measure of the size spread. Such an estimate of the size spread is difficult to obtain by other means for certain systems, e.g., microemulsions. [Another technique, for obtaining size distribution, is discussed in the papers of Provencher (1982), Stelzer and Ruf (1983) and Stelzer, Ruf and Grell (1982) cited in Section D below.]

Systematic examinations of the effects of polydispersity have not been carried out experimentally, but the paper of Gaylor, Snook and van Megen (1981) cited earlier in this section has verified the equations of Pusey (1980) and Weissman (1980) using Brownian dynamics computer simulations on dispersions interacting through only colloidal forces (i.e., without hydrodynamic interactions). Just recently, Kotlarchyk and Chen (1983) have presented a method for analyzing small-angle neutron scattering (SANS) data from interacting, polydisperse systems.

D. Experimental Techniques

Numerous experimental techniques have been developed over the years to study various aspects of colloidal dispersions. These include techniques for measuring specific equilibrium properties [e.g., osmotic pressures; see Ottewill (1977), Homola and Robertson (1976), Homola, Snook and van Megen (1977), Pieranski (1983), and Dickinson (1983)] or dynamic properties [e.g., rheological parameters; see discussions in Chapter 7, Section C and references therein; see, also, Goodwin (1983) for a description of shear-flow, oscillatory flow and extensional-flow techniques]. However, a systematic analysis of the interaction effects on the macroscopic behavior of interacting dispersions requires, in addition to equilibrium and transport properties, information on the local (static and dynamic) structure of the dispersions (obtained, preferably, by non-intrusive techniques). This has been made possible recently, due to significant advances in radiation scattering techniques [such as quasi-elastic light scattering (QELS), small-angle neutron scattering (SANS), small-angle X-ray scattering (SAXS), and neutron spin-echo spectrometry (NSES)]. Since the primary focus of this Chapter has been on the equilibrium and dynamic aspects of the local geometric structure of dispersions, this present summary of experimental techniques will be confined to the above radiation scattering techniques, which are being developed and refined specifically to provide a direct access to such information. We have already referred to numerous specific examples of the use of these experimental methods in the previous sections of this Chapter. Therefore, the present discussion will be confined to a brief introduction to the advantages of each technique, a summary of the types of information one usually needs, and an up-to-date list of monographs, books and research papers that discuss the various aspects of instrumentation and applications.

The advantage of having direct access to the local structure (through the static structure factors and the intermediate scattering functions) is considerable. This implies that one has the opportunity to examine and interpret the interaction effects more directly, with fewer assumptions and approximations than would be needed if one has access to only the average macroscopic properties (such as viscosity, yield stress or osmotic pressure). Obviously, a combination of both types of experimental data is usually the most desirable (see, for example, the discussion near the end of Section B.3 in this Chapter). Many of the theoretical concepts and analytical procedures needed for interpreting scattering data have been described already, or are implicit in the discussion presented, in the previous sections of this Chapter. We illustrate this by summarizing the types of information one usually obtains from laser light scattering measurements and how these are related to some of the structural details and properties of the dispersion.

Static and dynamic light scattering techniques are well-established methods for investigating supramolecular dynamics, and excellent treatments of the basic concepts of both the theoretical and the practical aspects of this technique, including information on various experimental options, are available in the literature [see, for instance, Kerker (1969); Cummins and Pike (1974); Berne and Pecora (1976); and Degiorgio, Corti and Giglio (1980)]. The light scattering measurements required for the investigation of interaction effects fall under two broad classes:

a. Equilibrium (static) structure factor measurements, obtained from the average intensity of the scattered light as a function of scattering angle, and

b. Dynamic measurements, derived from the temporal decay of the intensity autocorrelation function.

The latter measurements lead to the determination of

1. the single-particle diffusion coefficient, which describes the unhindered Brownian motion of a typical particle,

2. the mutual diffusion coefficient, which accounts for the collective influence of particle interactions and represents the decay of local concentration fluctuations, and

3. the self-diffusion coefficient, which is effectively the tracer diffusion coefficient of a typical particle in a cloud of its neighbors.

Precisely which of these is measured from the decay of intensity autocorrelation depends on the angle of measurement and the time interval over which the decay is monitored (see Figure 6.D.1). The reasons for these, and detailed discussions of the necessary theories, are available in the references cited above. The determination of the static structure factor from the intensity measurements is also described in these references [see, in addition, Pusey (1982)], and the recent book by Dahneke (1983) contains numerous current uses of QELS for a variety of colloidal systems. The literature on applications of QELS to monodispersed and polydispersed model colloids is summarized in Dickinson (1983). Most of the initial thrust in the development and interpretation of scattering techniques was on refining experimental procedures and measuring mutual and self-diffusion coefficients (to obtain average size, size spread, and charge of the particles) for a wide variety of colloidal materials such as microemulsions, vesicles, and micellar solutions [see Dahneke (1983); Degiorgio, Corti and Giglio (1980); Chen, Chu and Nossal (1981)]. Of late, more sophisticated analyses have become possible; notable among these are the development of structural models for micelles by explicit introduction of interaction effects in the structure factor (see Section B.2) and development of procedures for inverting the intermediate scattering functions to obtain size distributions [Stelzer and Ruf (1983); Stelzer, Ruf and Grell (1982);

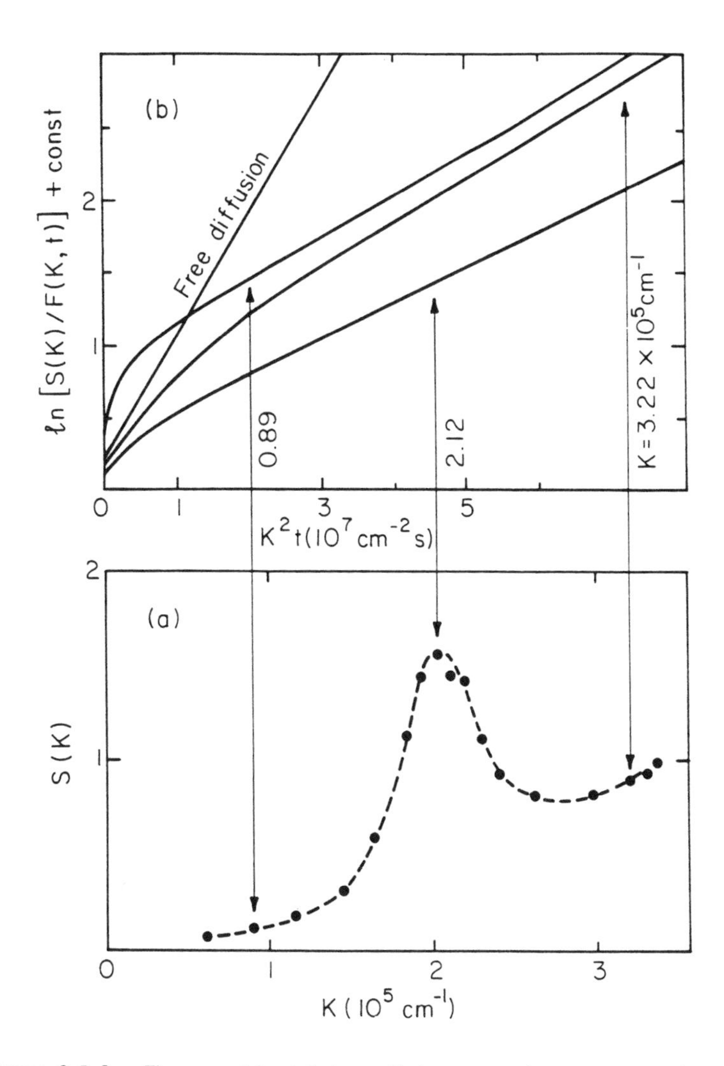

Figure 6.D.1 The normalized intermediate scattering functions (upper Figure) measured at various values of the scattering vector K [see Boon and Yip (1980) and Pusey (1978)].

The type of diffusion coefficient measured (i.e., mutual or self) depends on the K-value at which the scattering function is measured.

Provencher (1982)].

Laser light scattering is useful for particle sizes and inter-particle separations of the order of the wavelength of the monochromatic source used. For smaller species such as micelles and microemulsions, small-angle neutron scattering (SANS) can be used in a manner analogous to that of QELS. Ottewill (1982) presents an excellent introductory treatment of SANS, its differences from QELS, its strengths and its advantages. Some illustrative examples of the uses of SANS, such as particle size determination and measurement of thickness of steric layers, are also available in this paper. Other overviews and specific applications (e.g., determination of _intra_-micellar structure and critical behavior of microemulsions) are presented in Hayter and Penfold (1983), Bendedouch, Chen and Koehler (1983a,b), Bendedouch and Chen (1983a,b), and Kotlarchyk, Chen and Huang (1982, 1983). The theoretical ideas described in these papers are applicable to small-angle X-ray scattering (SAXS) as well. Some additional examples on using SAXS to study ordered and disordered dispersions and other related references can be found in Ise et al. (1980, 1983).

A relatively newer technique that offers considerable potential in the investigations of interest here is an extension of neutron scattering known as neutron spin-echo spectrometry (NSES). NSES measures the same physical quantity as QELS, but at generally larger momentum transfers (i.e., scattering vectors) and at smaller times than the latter. The experimental study of Hayter and Penfold (1981) described in Section B.2 employs NSES. Additional information on the basic concepts and instrumental details can be obtained from Mezei (1972), Richter et al. (1978), Hayter (1978), and Hayter and Penfold (1979, 1983).

Finally, a multidetector scattering technique, which measures the cross correlation of the intensity fluctuations of light scattered by the colloidal fluid through two different wave vectors, has been developed recently by Clark, Ackerson and Hurd (1983). This technique, called cross correlation intensity fluctuation spectroscopy (CCIFS), uses two detectors and focuses on a small scattering volume comparable to the range of positional correlation in the fluid. The above paper reports some initial data on dispersions of polystyrene latex particles of diameter 0.109 μm. Clark, Ackerson and Hurd also suggest that this procedure may be extended to X-ray scattering as well.

Innovative uses and extensions of the techniques listed in this section are being reported in the literature more frequently than in the past, and the resulting increases in the quality and the depth of experimental data on concentrated dispersions are expected to open new avenues in the study of these and other disordered and ordered materials.

Research Needs

As evident from the discussions presented in this Chapter, research on concentrated or strongly-interacting dispersions has grown rapidly in the last five years. The growing interest in this area can be traced directly to the fact that dispersions encountered in most of the practical applications are often far from dilute, but the momentum for the rapid growth in research has come from advances in the theories of classical many-body systems. As mentioned in Chapter 1, this cross-disciplinary interaction also has potential impact on the study of molecular systems as well. Despite the above surge in activity, there are many unresolved problems that need further investigation. The recent developments in theoretical and experimental procedures offer the means for a systematic approach to these problems.

- Additional work on order/disorder transitions in model colloids is needed. Of particular interest are order formation in dilute systems and ordering in binary and multicomponent model colloids.
- Additional experimental measurements of interparticle spacing in dilute, ordered dispersions are needed to study the nature of interaction forces in strongly-interacting dispersions.
- Experimental work on shear-induced melting and structural transitions (using cross correlation intensity fluctuation measurements and other multiprobe techniques) promises to lead to a better understanding of growth of clusters and their distortion under shear. Similar experiments using binary colloidal crystals will be useful.
- The use of model colloids as model many-body systems (both two-dimensional and three-dimensional) deserves further study (see Section A.3).
- The role of attractive forces and weak repulsive forces in producing short-range and long-range ordering requires further theoretical investigation. Analytical theories for describing these phenomena are needed. Existing theories also need to be examined against the results of exact computer experiments.
- Theories of multicomponent colloidal dispersions are needed. Effects of polydispersity on mutual diffusion coefficients and on the formation of glassy amorphous structures need careful study.

- Experimental measurements of equilibrium properties of monodispersed and polydispersed suspensions of model colloids are needed to test the theories.

- Systematic studies of low-angle radiation scattering using dispersions of controlled polydispersity will be useful for developing methods for determining polydispersity of microemulsions and similar colloids.

- Direct inversion of the intermediate scattering function to obtain size distribution seems to be a viable procedure.

- Computer experiments using Brownian dynamics algorithms are theoretically possible, although computationally formidable. It will be very instructive to use such experiments to study various ways of including multiparticle hydrodynamic interactions in the analysis of many-body effects. Examination of 'hydrodynamic screening', as an alternative to pair-additive hydrodynamic interactions, deserves attention.

- Concentration effects and clustering of particles near walls are important in applications such as membrane filtration and cake filtration. Even a preliminary examination of these will be useful.

- As mentioned in Section D, newer techniques such as neutron spin-echo spectroscopy and cross correlation intensity fluctuation spectroscopy can probe structural details that cannot be obtained by more traditional techniques such as photon correlation spectroscopy. Experiments using these newer techniques on model colloids and micelles are needed.

- Preliminary examination of structure of flocs is within reach and will be of interest in the study of rheology of flocculated dispersions (see Chapters 4 and 7).

- Experiments using model colloids of non-spherical particles will be useful for understanding concentrated dispersions at large ionic strengths. At low ionic strengths (strongly-interacting dispersions), the effects of shape are likely to be less important (see Section B.4).

- Ordering in solutions of macroions and biocolloids is not well understood and additional experiments are needed.

REFERENCES: SECTION A. STATIC STRUCTURES: COLLOIDAL CRYSTALS

Ackerson, B. J. and Clark, N. A., Phys. Rev. Lett. 46, 123 (1981).

Ackerson, B. J. and Clark, N. A., Physica A 118, 221 (1983).

Alder, B. J. and Wainwright, T. E., Phys. Rev. 127, 359 (1962).

Barker, J. A., Lattice Theories of the Liquid State, Pergamon Press, London, U. K., 1963.

Barnes, C. J., Chan, D. Y. C., Everett, H. and Yates, D. E., J. Chem. Soc. Faraday Trans. II 74, 136 (1978).

Bernal, J. D. and Fankuchen, I., J. Gen. Physiol. 25, 111 (1941).

Benzing, D. W. and Russel, W. B., J. Colloid Interface Sci. 83, 178 (1981).

Brenner, S. L., J. Phys. Chem. 80, 1473 (1976).

Care, C. M. and March, N. H., Adv. Phys. 24, 101 (1975).

Castillo, C. A., A Monte Carlo Study of Order/Disorder Transitions in Colloidal Dispersions, M. S. Thesis, Rensselaer Polytechnic Institute, Troy, NY, 1982.

Castillo, C. A. and Rajagopalan, R., Thermodynamic Properties of Interacting Dispersions: Hard-Sphere and Perturbation Theories, Paper presented at the 29th IUPAC Congress, Cologne, F.R.G., June, 1983.

Castillo, C. A., Rajagopalan, R. and Hirtzel, C. S., Rev. in Chem. Eng., to appear (1984).

Chaikin, P. M., Pincus, P., Alexander, S. and Hone, D., J. Colloid Interface Sci. 89, 555 (1982).

Clark, N. A., Hurd, A. J. and Ackerson, B. J., Nature 281, 57 (1979).

Efremov, I. F., pp. 85-192 in Matijević (1976).

Feynman, R. P., Leighton, R. B. and Sands, M., The Feynman Lectures on Physics, Vol. 1, Addison-Wesley, Reading, Mass., 1963.

Fitch, R. M. (ed.), Polymer Colloids, Plenum, New York, NY, 1971.

Fitch, R. M. (ed.), Polymer Colloids II, Plenum, New York, NY, 1980.

Forsyth, P. A., Marcelja, J. S., Mitchell, D. J. and Ninham, B. W., Adv. Colloid Interface Sci. 9, 37 (1978).

Furusawa, K. and Yamashita, S., J. Colloid Interface Sci. 89, 574 (1982).

Goldfinger, G. (ed.), Clean Surfaces, Marcel Dekker, New York, NY, 1970.

Hachisu, S. and Kobayashi, Y., J. Colloid Interface Sci. 46, 470 (1974).

Hachisu, S., Kobayashi, Y. and Kose, A., J. Colloid Interface Sci. 42, 342 (1973).

Hachisu, S., Kose, A., Kobayashi, Y. and Takano, K., J. Colloid Interface Sci. 55, 499 (1976).

Hachisu, S. and Takano, K., Adv. Colloid Interface Sci. 16, 233 (1982).

Hachisu, S. and Yoshimura, S., Nature 283, 188 (1980).

Hastings, R., J. Chem. Phys. 68, 675 (1978a).

Hastings, R., Phys. Lett. A 67, 316 (1978b).

Hiltner, P. A. and Krieger, I. M., J. Phys. Chem. 73, 2386 (1969).

Hirtzel, C. S. and Rajagopalan, R., Naturwissenschaften 70, 569 (1983).

Imai, N., Influence of Thermal Fluctuation of Concentration-Potential in the Colloid Assembly on the Ordering Process, Paper presented at the 57th Colloid and Surface Science Symposium, Toronto, Canada, June 1983a.

Imai, N., Nagoya Univeristy, Japan; Personal Communication, July, 1983b.

Ise, N., pp. 115-132 in Yoshida and Ise (1983).

Ise, N. and Okubo, T., Accounts Chem. Res. 13, 303 (1980).

Ise, N., Okubo, T., Kitano, H., Sugimura, M. and Date, S., Naturwissenschaften 69, 544 (1982).

Ise, N., Okubo, T., Sugimura, M., Ito, K. and Nolte, H. J., J. Chem. Phys. 78, 536 (1983a).

Ise, N., Okubo, T., Yamamoto, K., Matsuoka, H., Kawai, H., Hashimoto, T. and Fujimura, M., J. Chem. Phys. 78, 541 (1983b).

Ise, N., Okubo, T., Yamamoto, K., Kawai, H., Hashimoto, T., Fujimura, M. and Hiragi, Y., J. Amer. Chem. Soc. 102, 7901 (1980).

Israelachvili, J. N. and Ninham, B. W., J. Colloid Interface Sci. 58, 14 (1977).

Joyce, M., Cell Structures in Colloidal Fluids, M. S. Thesis, Rensselaer Polytechnic Institute, Troy, NY, 1983.

Kirkwood, J. G., Chem. Rev. 19, 275 (1936).

Kirkwood, J. G., J. Chem. Phys. 7, 919 (1939).

Kirkwood, J. G. and Mazur, J., J. Polym. Sci. 9, 519 (1952).

Kose, A. and Hachisu, S., J. Colloid Interface Sci. 46, 460 (1974).

Kose, A., Ozaki, M., Takano, K., Kobayashi, Y. and Hachisu, S., J. Colloid Interface Sci. 44, 330 (1973).

Krieger, I. M. and Hiltner, P. A., pp. 63-72 in Fitch (1971).

Krumrine, P. R. and Vanderhoff, J. W., pp. 289-312 in Fitch (1980).

Langmuir, I., J. Chem. Phys. 6, 873 (1938).

Levine, S., Proc. Roy. Soc. London A 170, 165 (1939).

Lindemann, F. A., Z. Phys. 11, 609 (1910).

Maeda, Y. and Hachisu, S., Colloids Surfaces 6, 1 (1983).

Marcelja, S., Mitchell, D. J. and Ninham, D. W., Chem. Phys. Lett. 43, 353 (1976).

March, N. H. and Tosi, M. P., Phys. Lett. A 50, 224 (1974).

Matijević, E. (ed.), Surface and Colloid Science, Vol. 8, Wiley-Interscience, New York, NY, 1976.

Matijević, E. (ed.), Surface and Colloid Science, Plenum, New York, NY, Vol. to appear, 1984.

Münster, A., Statistical Thermodynamics, Vol. 2, Springer-Verlag, Berlin, F.R.G., 1974.

Onsager, L., Ann. N. Y. Acad. Sci. 51, 627 (1949).

Pieranski, P., Contemp. Phys. 24, 25 (1983).

Rajagopalan, R. and Hirtzel, C. S., Statistical Mechanical Theories of Colloidal Fluids, to appear in Matijević (1984).

Russell, W. B. and Benzing, D. W., J. Colloid Interface Sci. 83, 163 (1981).

Snook, I. and van Megen, W., J. Chem. Soc. Faraday Trans. II 72, 216 (1976a).

Snook, I. and van Megen, W., J. Colloid Interface Sci. 57, 47 (1976b).

Sogami, I., Phys. Lett. A 96, 199 (1983).

Sogami, I., Miyoshi, S., Yoshiyama, T. and Ise, N., Structure of Crystalline Ordering of Spherical Macroions in Dilute Dispersions, to appear (1984).

Stigter, D., Rec. Trav. Chim. Pays-Bas. 73, 593 (1954).

Thomsen, D. E., Science News 121, 334 (1982).

Vanderhoff, J. W., van den Hul, H. J., Tausk, R. J. and Overbeek, J. Th. G., pp. 15-44 in Goldfinger (1970).

van Megen, W. and Snook, I., J. Colloid Interface Sci. 53, 172 (1975a).

van Megen, W. and Snook, I., Chem. Phys. Lett. 35, 399 (1975b).

van Megen, W. and Snook, I., J. Colloid Interface Sci. 57, 40 (1976).

van Megen, W. and Snook, I., J. Chem. Phys. 66, 813 (1977).

van Megen, W. and Snook, I., Faraday Dicuss. Chem. Soc. 65, 92 (1978).

Wadati, M. and Toda, M., J. Phys. Soc. Japan 32, 1147 (1972).

Williams, R. C. and Smith, K., Nature 179, 119 (1957).

Williams, R., Crandall, R. S. and Wojtowics, P. J., Phys. Rev. Lett. 37, 348 (1976).

Yoshida, Z. I. and Ise, N. (eds.), Biomimetic Chemistry, Elsevier, Amsterdam, The Netherlands, 1983.

Yoshimura, S. and Hachisu, S., Progr. Colloid Polymer Sci. 68, 59 (1983).

Ziman, J. M., Models of Disorder, Cambridge Univ. Press, Cambridge, U.K., 1979.

REFERENCES: SECTION B. STATIC STRUCTURE: COLLOIDAL FLUIDS

Ackerson, B. J. and Clark, N. A., Physica A 118, 221 (1983).

Ailawadi, N. K., Phys. Rep. 57, 241 (1980).

Andersen, H. C., Chandler, D. and Weeks, J. D., Phys. Rev. A 4, 1597 (1971).

Andersen, H. C., Chandler, D. and Weeks, J. D., J. Chem. Phys. 56, 3812 (1972).

Ashcroft, N. W. and Lekner, J., Phys. Rev. 145, 83 (1966).

Barker, J. A. and Henderson, D., J. Chem. Phys. 47, 2856 (1967a).

Barker, J. A. and Henderson, D., J. Chem. Phys. 47, 4714 (1967b).

Barker, J. A. and Henderson, D., Accounts Chem. Res. 4, 303 (1971a).

Barker, J. A. and Henderson, D., Phys. Rev. A 4, 806 (1971b).

Barker, J. A. and Henderson, D., Rev. Mod. Phys. 48, 587 (1976).

Barker, J. A. and Henderson, D., Scientific American 245 (5), 130 (1981).

Barton, A. F. M., The Dynamic Liquid State, Longman, London, U. K., 1974.

Bendedouch, D. and Chen, S-H., J. Phys. Chem. 87, 1653 (1983).

Bendedouch, D., Chen, S-H. and Koehler, W. C., J. Phys. Chem. 87, 2621 (1983).

Beresford-Smith, B. and Chan, D. Y. C., Chem. Phys. Lett. 92, 474 (1982).

Beresford-Smith, B. and Chan, D. Y. C., Faraday Discuss. Chem. Soc. 76, 65 (1983).

Berne, B. J. (ed.), Statistical Mechanics, Part A, Plenum, New York, NY, 1977.

Beunen, B. A. and White, L. R., Colloids Surfaces 3, 371 (1981).

Blum, L. and Stell, G., J. Chem. Phys. 71, 42 (1979).

Brown, J. C., Pusey, P. N., Goodwin, J. W. and Ottewill, R. H., J. Phys. A: Math. Gen. 8, 664 (1975).

Castillo, C. A., Equilibrium Structure of Colloidal Dispersions, Ph.D. Dissertation, Rensselaer Polytechnic Institute, Troy, New York, NY, 1984.

Castillo, C. A. and Rajagopalan, R., Thermodynamic Properties of Interacting Dispersions: Hard-Sphere and Perturbation Theories, Paper presented at the 29th IUPAC Congress, Cologne, F.R.G., June, 1983.

Castillo, C. A., Rajagopalan, R. and Hirtzel, C. S., Rev. in Chem. Eng., in press (1984).

Chandler, D., Weeks, J. D. and Andersen, H. C., Science 220, 787 (1983).

Clark, N. A. and Ackerson, B. J., Phys. Rev. Lett. 44, 1005 (1980).

Cowell, C. and Vincent, B., J. Colloid Interface Sci. 87, 518 (1982).

Croxton, C. A., Liquid State Physics: A Statistical Mechanical Introduction, Cambridge Univ. Press, London, U. K., 1974.

Dickinson, E., Chem. Phys. Lett. 57, 148 (1978).

Dickinson, E., J. Chem. Soc. Faraday Trans. II 75, 466 (1979).

Dickinson, E., pp. 150-179 in Everett (1983).

Dickinson, E., Parker, R. and Lal, M., Chem. Phys. Lett. 79, 578 (1981).

Everett, D. H. (ed.), Specialist Periodical Reports, Colloid Science, Vol. 4, The Royal Soc. of Chem., London, U. K., 1983.

Grimson, M. J., J. Chem. Soc. Faraday Trans. II 79, 817 (1983).

Hansen, J-P., Phys. Rev. A 8, 3096 (1973).

Hansen, J-P. and Hayter, J. B., Mol. Phys. 46, 651 (1982).

Hansen, J-P. and McDonald, I. R., Theory of Simple Liquids, Academic Press, New York, NY, 1976.

Harris, S., J. Phys. A 9, 1895 (1975).

Hayter, J. B., Faraday Discuss. Chem. Soc. 76, 7 (1983).

Hayter, J. B. and Penfold, J., J. Chem. Soc. Faraday Trans. I 77, 1851 (1981a).

Hayter, J. B. and Penfold, J., Mol. Phys. 42, 109 (1981b).

Hayter, J. B. and Zulauf, M., Colloid Polymer Sci. 260, 1023 (1982).

Henderson, D. and Barker, J. A., Phys. Rev. A 1, 1266 (1970).

Hirtzel, C. S. and Rajagopalan, R., Naturwissenschaften 70, 569 (1983).

Keavey, R. P. and Richmond, P., J. Chem. Soc. Faraday II 72, 773 (1976).

Kirkwood, J. G. and Mazur, J., J. Polym. Sci. 9, 519 (1952).

Lebowitz, J. L. and Percus, J. K., Phys. Rev. 144, 251 (1966).

Long, J. A., Osmond, D. W. J. and Vincent, B., J. Colloid Interface Sci. 42, 545 (1973).

Maleki, S., Hirtzel, C. S. and Rajagopalan, R., Phys. Lett. A 97, 289 (1983).

McQuarrie, D. A., Statistical Mechanics, Harper and Row, New York, NY, 1976.

Metropolis, N., Rosenbluth, A. W., Rosenbluth, M. N., Teller, A. H. and Teller, E., J. Chem. Phys. 21, 1087 (1953).

Münster, A., Statistical Thermodynamics, Vol. 2, Springer-Verlag, Berlin, F.R.G., 1974.

Ng, K.-C., J. Chem. Phys. 61, 2680 (1974).

Nieuwenhuis, E. A. and Vrij, A., J. Colloid Interface Sci. 72, 321 (1979).

Palmer, R. G. and Weeks, J. D., J. Chem. Phys. 58, 4171 (1973).

Reichl, L. E., A Modern Course in Statistical Physics, Univ. of Texas Press, Austin, Texas, 1980.

Salacuse, J. J. and Stell, G., J. Chem. Phys. 77, 3714 (1982).

Schaefer, D. W., J. Chem. Phys. 66, 3980 (1977).

Stephen, M. J., J. Chem. Phys. 55, 3878 (1971).

Valleau, J. P. and Torrie, G. M., pp. 169-194 in Berne (1977).

van Beurten, P. and Vrij, A., J. Chem. Phys. 74, 2744 (1981).

van Megen, W. and Snook, I., Faraday Discuss. Chem. Soc. 65, 92 (1978).

Vrij, A., Chem. Phys. Lett. 53, 144 (1978a).

Vrij, A., J. Chem. Phys. 69, 1742 (1978b).

Vrij, A., J. Chem. Phys. 71, 3267 (1979).

Vrij, A., Nieuwenhuis, E. A., Fijnaut, H. M. and Agterof, W. G. M., Faraday Discuss. Chem. Soc. 65, 101 (1978).

Watts, R. O. and McGee, I. J., Liquid State Chemical Physics, Wiley-Interscience, New York, NY, 1976.

REFERENCES: SECTION C. DYNAMIC STRUCTURE OF INTERACTING DISPERSIONS

Ackerson, B. J., J. Chem. Phys. 64, 242 (1976).

Ackerson, B. J., Pusey, P. N. and Tough, R. J. A., J. Chem. Phys. 76, 1279 (1982).

Alder, B. J. and Wainwright, T. E., J. Chem. Phys. 27, 1208 (1957).

Bacon, J., Dickinson, E. and Parker, R., J. Chem. Soc. Faraday Trans. II 79, 91 (1983a).

Bacon, J., Dickinson, E. and Parker, R., Faraday Discuss. Chem. Soc. 76, 165 (1983b).

Barton, A. F. M., The Dynamic Liquid State, Longman, London, U. K., 1974.

Batchelor, G. K., J. Fluid Mech. 74, 1 (1976).

Boon, J. P. and Yip, S., Molecular Hydrodynamics, McGraw-Hill, New York, NY, 1980.

Brown, J. C., Pusey, P. N., Goodwin, J. W. and Ottewill, R. H., J. Phys. A.: Math. Gen. 8, 664 (1975).

de Gennes, P. G., Physica (Utrecht) A 25, 825 (1959).

Degiorgio, V., Corti, M. and Giglio, M. (eds.), Light Scattering in Liquids and Macromolecular Solutions, Plenum, New York, NY, 1980.

Deutch, J. M. and Oppenheim, I., J. Chem. Phys. 54, 3547 (1971).

Dickinson, E., pp. 150-179 in Everett (1983).

Ermak, D. L., J. Chem. Phys. 62, 4189 (1975a).

Ermak, D. L., J. Chem. Phys. 62, 4197 (1975b).

Ermak, D. L. and McCammon, J. A., J. Chem. Phys. 69, 1352 (1978).

Ermak, D. L. and Yeh, Y., Chem. Phys. Lett. 24, 243 (1974).

Everett, D. H. (ed.), Specialist Periodical Reports, Colloid Science, Vol. 4, The Royal Soc. Chem., London, U. K., 1983.

Felderhof, B. U., Physica (Utrecht) A 89, 373 (1977).

Felderhof, B. U., J. Phys. A 11, 929 (1978).

Gaylor, K. J., Snook, I. and van Megen, W., J. Chem. Phys. 75, 1682 (1981).

Gaylor, K. J., Snook, I., van Megen, W. and Watts, R. O., Chem. Phys. 43, 233 (1979).

Gaylor, K. J., Snook, I., van Megen, W. and Watts, R. O., J. Phys. A 13, 2513 (1980).

Kotlarchyk, M. and Chen, S-H., J. Chem. Phys. 79, 2461 (1983).

Mazur, P. and van Saarloos, W., Physica (Utrecht) A 115, 21 (1982).

Murphy, T. J. and Aguirre, J. L., J. Chem. Phys. 57, 2098 (1972).

Nijboer, B. R. A. and Rahman, A., Physica 32, 415 (1966).

Phillies, G. D. J., J. Chem. Phys. 67, 4690 (1977).

Pusey, P. N., J. Phys. A: Math. Gen. 8, 1433 (1975).

Pusey, P. N., J. Phys. A: Math. Gen. 11, 119 (1978).

Pusey, P. N., Phil. Trans. Roy. Soc. London A 293, 429 (1979).

Pusey, P. N., pp. 1-29 in Degiorgio, Corti and Giglio (1980).

Pusey, P. N. and Tough, R. J. A., J. Phys. A: Math. Gen. 15, 1291 (1982).

Pusey, P. N. and van Megen, W., J. Phys. (Paris) 44, 285 (1983).

Rahman, A., Phys. Rev. A 136, 404 (1964).

Rajagopalan, R., The Statistical Mechanics of Dispersion Dynamics: The Colloid As a Grainy Fluid, Paper presented at the AIChE Annual Meeting, Los Angeles, CA, Nov., 1982.

Rotne, J. and Prager, S., J. Chem. Phys. 50, 4831 (1969).

Snook, I., van Megen, W. and Tough, R. J. A., J. Chem. Phys. 78, 5825 (1983).

Tsang, T., Physica (Utrecht) A 96, 359 (1979).

Tsang, T. and Jenkins, W. D., Mol. Phys. 41, 797 (1980).

Tsang, T. and Tang, H. T., J. Chem. Phys. 76, 3873 (1982).

Tutt, S., Brownian Dynamics Simulation of Colloidal Dynamics, M. S. Thesis, Rensselaer Polytechnic Institute, Troy, NY, 1982.

van Megen, W., Snook, I. and Pusey, P. N., J. Chem. Phys. 78, 931 (1983).

Venkatesan, M., Structure and Dynamics of Colloidal Dispersions, Ph.D. Dissertation, Rensselaer Polytechnic Institute, Troy, NY, 1984.

Venkatesan, M., Hirtzel, C. S., and Rajagopalan, R., Effect of Many-Body Interactions on the Self-diffusion Coefficients of Colloidal Particles in Concentrated Dispersions, Paper presented at the 57th ACS Colloid and Surface Science Symposium, Toronto, Canada, June 1983.

Vineyard, G. H., Phys. Rev. 110, 999 (1958).

Weissman, M. B., J. Chem. Phys. 72, 231 (1980).

Weissman, M. B., Pan. R. C. and Ware, B. R., J. Chem. Phys. 68, 5069 (1978).

Yamakawa, H., Modern Theory of Polymer Solutions, Harper and Row, New York, NY, 1971.

REFERENCES: SECTION D. EXPERIMENTAL TECHNIQUES

Bendedouch, D. and Chen, S-H., J. Phys. Chem. 87, 1473 (1983a).

Bendedouch, D. and Chen, S-H., J. Phys. Chem. 87, 1653 (1983b).

Bendedouch, D., Chen, S-H. and Koehler, W. C., J. Phys. Chem. 87, 153 (1983a).

Bendedouch, D., Chen, S-H. and Koehler, W. C., J. Phys. Chem. 87, 2621 (1983b).

Berne, B. J. and Pecora, R., Dynamic Light Scattering with Applications to Chemistry, Biology and Physics, Wiley, New York, NY, 1976.

Boon, J. P. and Yip, S., Molecular Hydrodynamics, McGraw-Hill, New York, NY, 1980.

Chen, S-H., Chu, B. and Nossal, R. (eds.), Scattering Techniques Applied to Supramolecular and Non-equilibrium Systems, Plenum, New York, NY, 1981.

Clark, N. A., Ackerson, B. J. and Hurd, A. J., Multidetector Scattering As A Probe of Local Structure in Disordered Phases, Preprint; to be published; Ackerson, B. J., Personal Communication, August, 1983.

Cummins, H. Z. and Pike, E. R. (eds.), Photon Correlation and Light Beating Spectroscopy, Plenum, New York, NY, 1974.

Dachs, H. (ed.), Neutron Diffraction, Springer-Verlag, Berlin, F.R.G., 1978.

Dahneke, B. E. (ed.), Measurement of Suspended Particles by Quasi-Elastic Light Scattering, Wiley, New York, NY, 1983.

Degiorgio, V., Corti, M. and Giglio, M. (eds.), Light Scattering in Liquids and Macromolecular Solutions, Plenum, New York, NY, 1980.

Dickinson, E., pp. 150-179 in Everett (1983).

Everett, D. H. (ed.), Specialist Periodical Reports, Colloid Science, Vol. 4, The Royal Soc. Chem., London, U. K., 1983.

Goodwin, J. W. (ed.), Colloidal Dispersions, The Royal Soc. Chem., London, U. K., 1982.

Goodwin, J. W., pp. 552-569 in Poehlein, Ottewill and Goodwin (1983).

Hayter, J. B., pp. 41-69 in Dachs (1978).

Hayter, J. B. and Penfold, J., Z. Phys. B 35, 199 (1979).

Hayter, J. B. and Penfold, J., J. Chem. Soc. Faraday Trans. I 77, 1851 (1981).

Hayter, J. B. and Penfold, J., Colloid Polymer Sci., to appear (1983).

Homola, A. and Robertson, A. A., J. Colloid Interface Sci. 54, 286 (1976).

Homola, A., Snook, I. and van Megen, W., J. Colloid Interface Sci. 61, 493 (1977).

Ise, N., Okubo, T., Yamamoto, K., Kawai, H., Hashimoto, T., Fujimura, M. and Hiragi, Y., J. Amer. Chem. Soc. 102, 7901 (1980).

Ise, N., Okubo, T., Yamamoto, K., Matsuoka, H., Kawai, H., Hashimoto, T. and Fujimura, M., J. Chem. Phys. 78, 541 (1983).

Kerker, M., The Scattering of Light and Other Electromagnetic Radiation, Academic Press, New York, NY, 1969.

Kerker, M., Zettlemoyer, A. C. and Rowell, R. L. (eds.) Colloid and Interface Science, Vol. 1, Academic Press, New York, NY, 1977.

Kotlarchyk, M., Chen, S-H. and Huang, J. S., J. Phys. Chem. 86, 3273 (1982).

Kotlarchyk, M., Chen, S-H. and Huang, J. S., Phys. Rev. A 28, 508 (1983).

Mezei, F., Z. Phys. 255, 146 (1972).

Ottewill, R. H., pp. 379-395 in Kerker, Zettlemoyer and Rowell (1977).

Ottewill, R. H., pp. 143-163 in Goodwin (1982).

Pieranski, P., Contemp. Phys. 24, 25 (1983).

Poehlein, G. W., Ottewill, R. H. and Goodwin, J. W. (eds.), Science and Technology of Polymer Colloids: Characterization, Stabilization and Application Properties, Vol. II, Martinus Nijhoff, The Hague, The Netherlands, 1983.

Provencher, S. W., Comp. Phys. Comm. 27, 213 (1982).

Pusey, P. N., J. Phys. A: Math. Gen. 11, 119 (1978).

Pusey, P. N., pp. 129-142 in Goodwin (1982).

Richter, D., Hayter, J. B., Mezei, F. and Ewen, B., Phys. Rev. Lett. 41, 1484 (1978).

Schulz-DuBois, E. O. (ed.), Photon Correlation Techniques in Fluid Mechanics, Springer-Verlag, Berlin, F.R.G., 1982.

Spach, G. (ed.), Physical Chemistry of Transmembrane Ion Motions, Elsevier, Amsterdam, The Netherlands, 1983.

Stelzer, E. and Ruf, H., pp. 37-43 in Spach (1983).

Stelzer, E., Ruf, H. and Grell, E., pp. 329-334 in Schulz-DuBois (1982).

7

Transport Properties of Interacting Dispersions

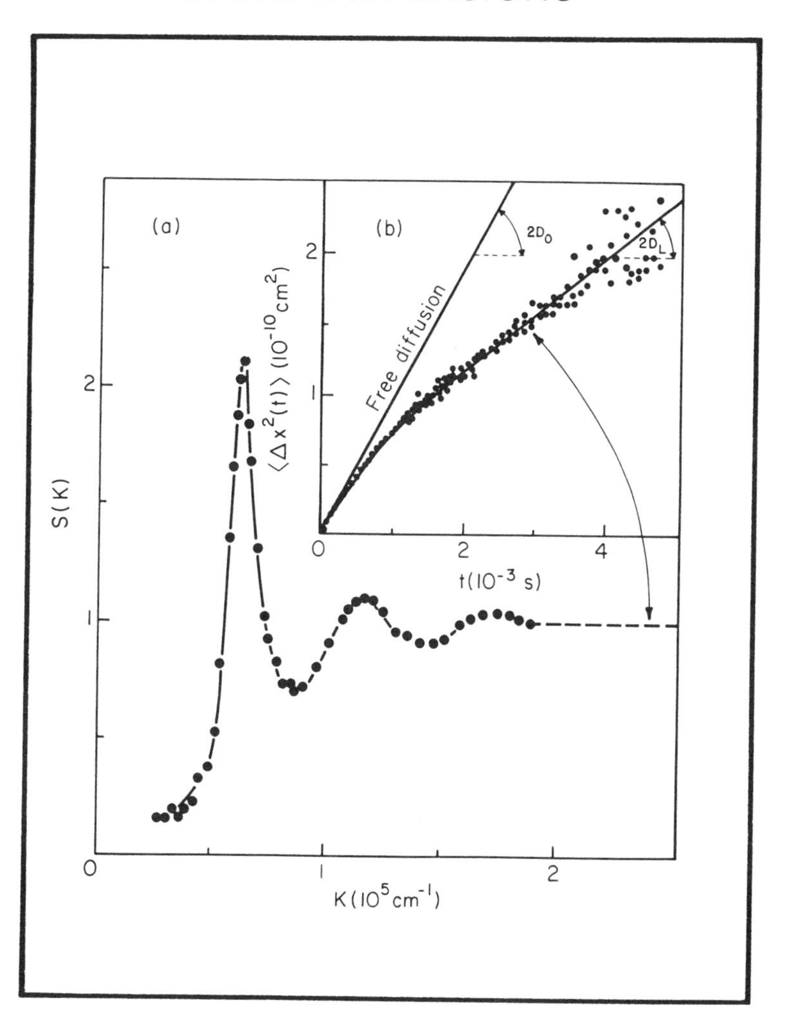

CHAPTER 7

TRANSPORT PROPERTIES OF INTERACTING DISPERSIONS

Overview

As emphasized in the discussion on the dynamic structure of dispersions in the previous Chapter, fundamental differences appear in the analysis of concentration effects on transport properties because of the complex role played by long-range hydrodynamic stress disturbances in this case. Firstly, the analytical and computational difficulties increase significantly because of the need for considering many-body interactions more rigorously. In addition, at least two other complications arise. One of these results from the competition between the motion of the fluid (which causes distortion of the electrical double layer or the steric layer) and the reaction of the stabilizing moeities to this distortion. This phenomenon is the cause of the electroviscous effects discussed in Section C of this Chapter (under Rheology) and may be significant at low shear rates. The other complication due to bulk flow arises from the distortion of the equilibrium radial distribution function of the particles by the flow field. This implies that computations of average macroscopic properties, such as sedimentation coefficients, through ensemble averages of local properties may require probability distributions derived from the solution of appropriate Fokker-Planck equations (see, for instance, the discussion in the second half of Section B). The equilibrium Boltzmann distributions discussed in Chapter 6 (Section B) may not suffice.

The discussion in this Chapter will be restricted to three classes of phenomena, namely, diffusion, sedimentation and rheology, that reflect directly the mechanical behavior of dispersions and its dependence on colloidal interactions. Thus, other properties such as thermal diffusivity have been excluded. The properties and problems chosen for discussion here are based on the level of current interest in the literature.

The presentation below starts with a discussion of diffusion coefficients in concentrated dispersions since this forms a logical extension of the material presented in the last two sections of the previous Chapter. Furthermore, the interplay between bulk flow and colloidal interactions in this case is relatively simple compared to that in sedimentation and rheology. Of the last two, which are discussed in Sections B and C, rheology has received considerable attention in the literature and, hence, forms a sizeable part of the present Chapter.

A. Diffusion Coefficients

The examination of the diffusion properties of concentrated dispersions offers a direct method for studying the dynamical aspects of colloidal interactions because of the availability of non-intrusive probes for measuring diffusion coefficients and space-time correlation functions (see Figures 7.A.1 and 2; see also Section D of Chapter 6). As also discussed in the previous Chapter, two types of diffusion coefficients can be identified, and each contains information on hydrodynamic and colloidal interactions. The collective diffusion coefficient, measured at the low-scattering-vector limit in, say, quasi-elastic light scattering spectroscopy, is a measure of the relaxation of concentration fluctuations and provides the same information obtained in macroscopic gradient diffusion experiments. It can be written in terms of an osmotic driving force and a friction coefficient using what is usually known as 'generalized' Stokes-Einstein equation [see Pusey and Tough (1982); Hess (1980); Hanna, Hess and Klein (1982); see also Section C.2 of Chapter 6 and the references therein]. The collective diffusion coefficient thus depends on both the equilibrium structure and the hydrodynamic interactions. The self-diffusion coefficient (also known as the tracer diffusion coefficient), usually measured at large scattering vectors in dynamic light scattering, represents single-particle motion and is affected, in the short-time limit, only by the hydrodynamic interactions. At longer times ($\gtrsim 10^{-3}$s), colloidal forces may influence its magnitude further. The experimental techniques that have been used previously to measure self-diffusion coefficients have relied on radioactive labeling of particles [Wiersema, Wiersema and Overbeek (1969a,b); Kitchen, Preston and Wells (1976)] or dye

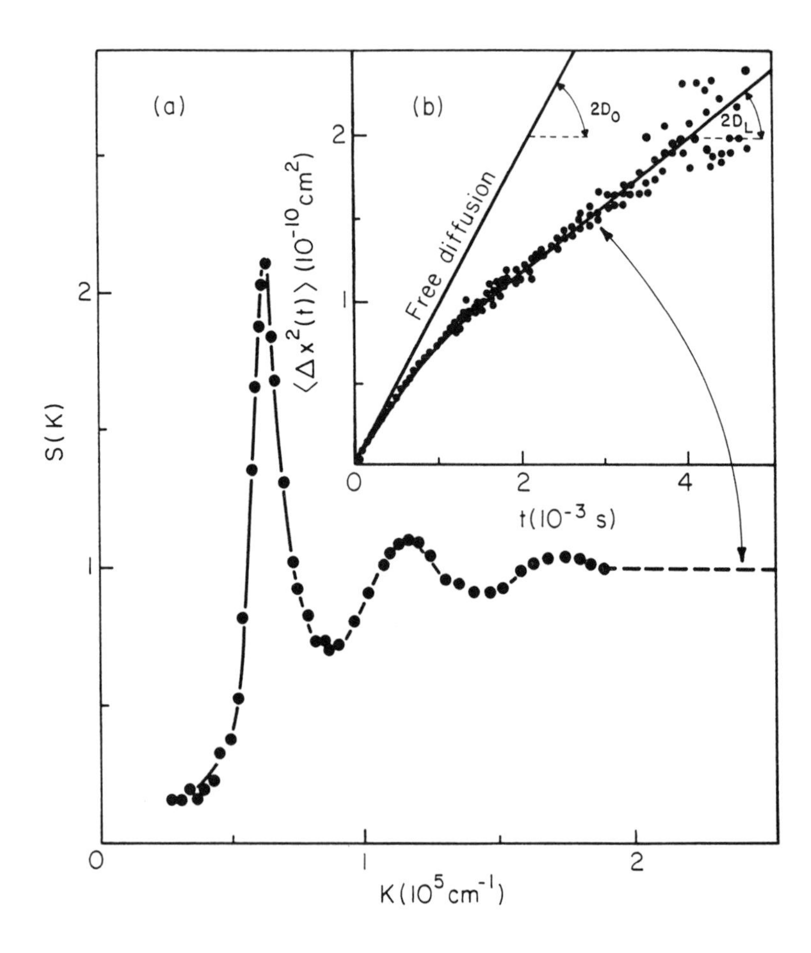

Figure 7.A.1 Mean-square displacements obtained from the intermediate scattering function at large scattering vectors [where $S(K) \rightarrow 1$]; see Boon and Yip (1980) and Pusey (1978).

The slope of the points shown (in the inset) is proportional to the long-time self-diffusion coefficient, D_L.

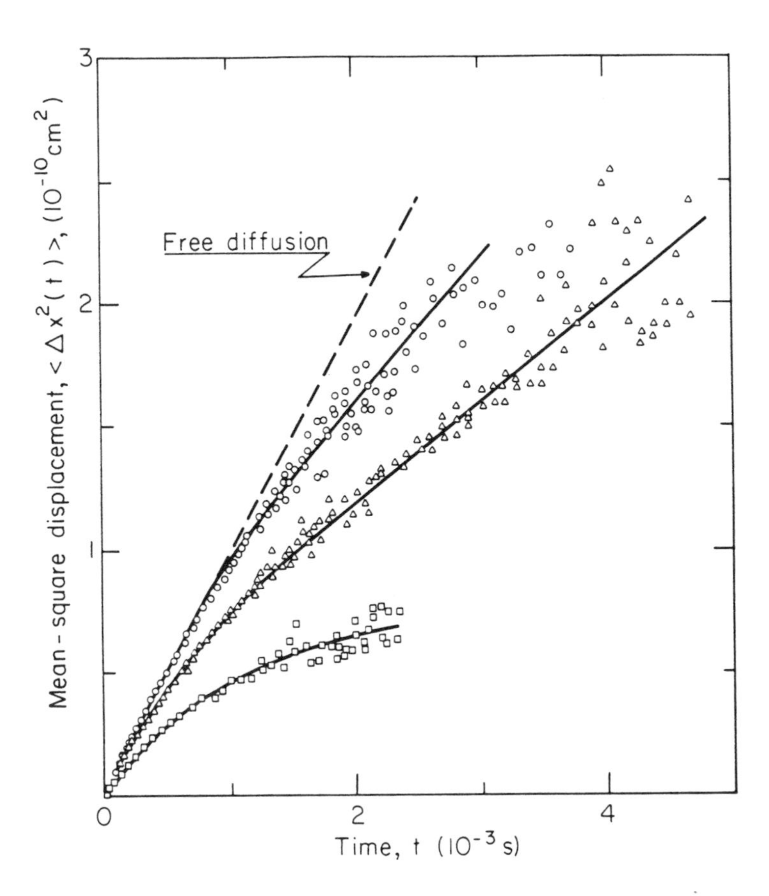

Figure 7.A.2 Mean-square displacements obtained using photon correlation spectroscopy as a function of concentration [from Pusey (1979)].

The self-diffusion coefficient at any instant (proportional to the slope of the line) decreases as concentration increases.

labeling techniques [Stigter and Mysels (1955); Stigter, Williams and Mysels (1955)]. These are being replaced slowly by light scattering and other radiation scattering techniques. While until recently only large-scattering-vector measurements of self-diffusion were possible, Kops-Werkhoven et al. (1982) have recently reported a novel 'tracer-particle' technique that can be used to measure self-diffusion coefficients at low scattering vectors (i.e., $K \to 0$). Physically, two different types of information are obtained in these two limits. Large-K measurements are affected only by the influence of short wavelength hydrodynamic disturbances and therefore measure self-diffusion of a particle in an essentially stationary 'average' configuration. On the other hand, low-K measurements include the effects of the motion of the neighbors [see Pusey, Fijnaut and Vrij (1982)] and represent the 'long-time' limit. The procedure of Kops-Werkhoven et al. uses tracer particles with a refractive index which differs from that of the solvent by a small but measureable amount, thereby making the tracer particles 'visible'. The tracer particles diffuse among 'filler' particles whose refractive index is practically the same as that of the solvent; thus the filler particles are 'invisible', but play their part in the particle interactions. Additional information on these and related experimental procedures may be obtained from the above references and others cited in Section D of Chapter 6 [see also Kops-Werkhoven and Fijnaut (1982)]. Equations relating the diffusion coefficients to the intermediate scattering functions and related theoretical concepts are also discussed in the references cited in Section C.2 of Chapter 6.

The major problems of current interest in the study of diffusion coefficients may be summarized as follows.

1. An understanding of the extent and influence of hydrodynamic interactions on self- and collective diffusion coefficients is required before the experimental data can be interpreted meaningfully [see, for example, Hess and Klein (1983); van Saarloos and Mazur (1983); Beenakker and Mazur (1982, 1983, 1984); and Beenakker (1984)].

2. Next follows the role of colloidal interactions themselves. In the case of collective diffusion coefficient, this is somewhat easier to handle since the equilibrium structure can be related to collective diffusion through the Stokes-Einstein relation.

3. What other information about the dispersion can be extracted from the measured diffusion coefficients is another question of significant practical interest.

4. Finally, the influence of polydispersity (of size, scattering power, etc.) is also an important factor that cannot be ignored without introducing inaccuracies in the interpretation of experimental data.

We review the significant pieces of information available in the literature on the above problems.

Hydrodynamic Interactions

A general discussion of diffusion coefficients in concentrated dispersions is presented in Hess (1980). Hess also discusses the memory-function formalism that can be used to relate the interactions to the observed diffusion coefficients [see, also, Hess and Klein (1983)]. The actual use of the available relations, however, normally requires approximations (e.g., for radial distribution functions) and assumptions (e.g., pair additivity of interaction potentials or hydrodynamic interactions) that may introduce considerable uncertainties in the calculated values. Some typical examples of these are available in Altenberger (1976), Felderhof (1978) and Ackerson (1976). Since statistical mechanical approximations such as truncation of virial series for the compressibility and hydrodynamic approximations such as using a particular expression for mobility compound each other, van Megen, Snook and Pusey (1983) recently chose to examine the effect of two-particle and three-particle mobilities on diffusion coefficients by focusing on <u>hard-sphere</u> suspensions. (The use of hard-sphere suspensions simplifies the computation of the equilibrium structure and eliminates the need to make further assumptions concerning the range of colloidal interactions in the computer simulations.) Both mutual (i.e., collective) diffusion coefficients and short-time self-diffusion coefficients have been computed using the two-particle and three-particle mobility expressions of Mazur and van Saarloos (1982). [The computations of van Megen, Snook and Pusey are more rigorous than those of Glendinning and Russel (1982).] The configurational mobility averaging needed in these computations has been carried out by direct Monte Carlo experiments and by integrating the mobilities using radial distribution functions presented by Verlet and Weis (1972). These computations seem to demonstrate that the two-particle mobility expression is sufficient for computing the short-time self-diffusion coefficient up to a volume fraction of about 0.3. However, in the case of collective diffusion coefficients, which are influenced by fluctuations over large separations, three- and possibly four-particle interactions seem to be necessary. It is also evident that at large volume fractions the computations are unreliable because of the uncertainties introduced by the range of hydrodynamic interactions. (See also the discussion on hydrodynamic interactions near the end of Section C.1 of Chapter 6.) In view of these practical difficulties, Snook, van Megen and Tough (1983) have attempted to derive, empirically, effective two-particle mobility tensors using experimentally-obtained values for the mutual and short-time self-diffusion coefficients. The mobility tensors are assumed to contain two screening constants to guarantee that the observed diffusion coefficients could be fitted sufficiently accurately over the entire range of volume fractions studied (up to about 0.5, the Kirkwood-Alder freezing point for hard-sphere suspensions). This procedure is analogous (in approach) to the determination of effective pair potentials of interactions from the equilibrium structure factor (see Section B.2 of Chapter 6), and its usefulness remains untested. It might be instructive to see if the effective mobilities computed in this manner can reproduce the intermediate scattering functions in a Brownian dynamics experiment.

Colloidal Interactions

In the case of monodispersed suspensions, the colloidal effects can be included in the above computations by using the appropriate radial distribution function and the osmotic pressure gradient. On the other hand, in the case of self-diffusion coefficients, only the short-time limit can be corrected for these effects. As evident from the experimental data shown in Figure 7.A.2, the diffusion coefficient decreases because of colloidal forces for times very much larger than the relaxation time for the initial, hydrodynamically-retarded Brownian motion. Although this trend can be computed using a Brownian dynamics experiment (see Figure 7.A.3), analytically there is not much known about this limit of the self-diffusion coefficient. At short times, the influence of colloidal forces on the self-diffusion coefficient enters indirectly through the rearrangement of the local geometric distribution of the particles. Venkatesan, Hirtzel and Rajagopalan (1983) have presented the change in the initial self-diffusion coefficients as a function of volume fraction (up to 0.15) for a few values

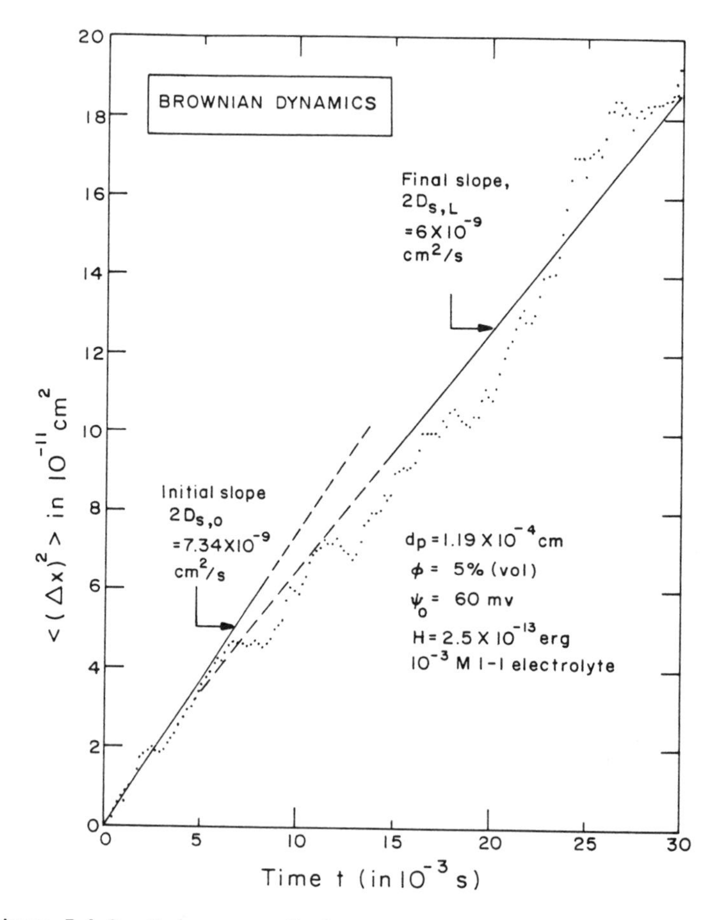

Figure 7.A.3 Mean-square displacements obtained using a Brownian dynamics experiment.

The final slope will be lower in this case because of colloidal interactions. The points shown are based on a preliminary computer experiment. The results will be smoother if a large enough number of averages is taken.

of ionic strength (see Figure 7.A.4), and it is evident that addition of electrolytes can increase the influence of hydrodynamic resistance beyond what is present in a hard-sphere dispersion unless the interaction forces are purely repulsive. In the presence of attraction, the increase in the nearest neighbors at large ionic strength is sufficient to increase the hydrodynamic resistance further. Results for long-time self-diffusion coefficients as functions of interaction parameters are also available from the solution of the two-particle Smoluchowski equation [see, for example, Venkatesan, Hirtzel and Rajagopalan (1984)]. Brownian dynamics experiments with hydrodynamic interactions have begun to appear in the literature only recently (see Section C.1 of Chapter 6), and this area of research appears promising.

Extraction of Physicochemical Parameters From Diffusion Coefficients

Stephen (1971) has demonstrated that the charge of the interacting particles can be determined from the measured mutual diffusion coefficient in a straightforward manner [see, also, Hess (1980)]. Doherty and Benedek (1974) have measured diffusion coefficients using dispersions of charged Bovine Serum Albumin, but find substantial deviations from Stephen's theory, particularly at low ionic strengths. Since Stephen did not consider hydrodynamic interactions in his derivation, correction for these may account for the observed discrepancies. In addition, the compressibility effects on the mutual diffusion coefficient must also be considered. Phillies (1974), in fact, has used the osmotic pressure gradient computed from the virial equation in the Stokes-Einstein relation to account for the inter-particle interactions (Stephen's theory considers particle-ion interactions, but not particle-particle interactions.) His expressions,

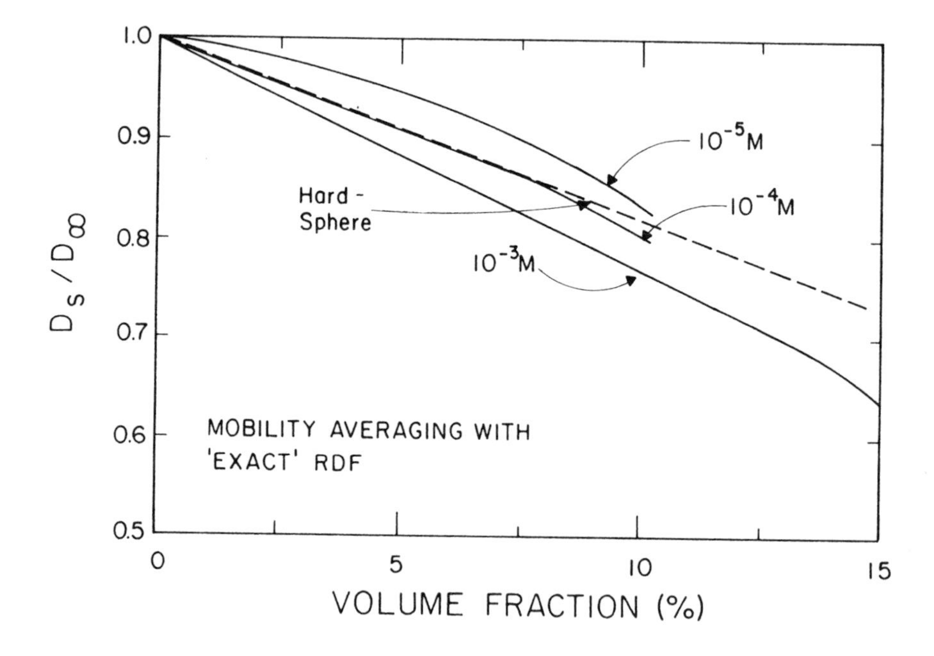

Figure 7.A.4 Normalized initial self-diffusion coefficient of a dispersion as a function of volume fraction [from Venkatesan, Hirtzel and Rajagopalan (1983)].

however, are limited to two special cases, namely, hard-sphere interactions at extreme dilution (i.e., without positional correlations in the structure) and dilute-gas radial distribution functions (which account for the pair interactions only). In view of this, the computations of Schor and Serrallach (1979), who use the theory of Phillies to examine the discrepancies reported by Doherty and Benedek, are unlikely to be sufficient. Nevertheless, it is clear that experimental measurements of diffusion coefficients can be used to extract information on at least some of the chemical parameters of the dispersions. Some recent work along these directions may be found in Corti and Degiorgio (1980), Nicoli and Dorshow (1983) and Dahneke (1983).

Polydispersity

Effects of polydispersity on diffusion coefficients have been described already in Section C.3 of Chapter 6. In addition to the papers cited there, a recent one by Pusey, Fijnaut and Vrij (1982) on polydispersed hard spheres deserves mention. This paper considers both size polydispersity and scattering-power polydispersity, although the discussion is limited to hard-sphere interactions only. Since size polydispersity leads to correlations between number fluctuations and concentration fluctuations (in addition to causing a distribution of scattering powers), Pusey, Fijnaut and Vrij show that the measured diffusion coefficients are only 'average' collective and self-diffusion coefficients. In contrast, in the case of scattering-power polydispersity, a rigorous analysis is shown to be possible and the two different decay modes (of the light scattering correlation function) are identified as the actual collective and self-diffusion coefficients. The results presented here are approximate in view of the simplifications made in the analysis, but the qualitative conclusions remain unaffected.

B. Sedimentation in Interacting Dispersions

Sedimentation of small particles (at low Reynolds numbers) has attracted the attention of fluid dynamicists for some time, although the colloidal aspects are only now being included in the analysis. There are some excellent treatments of expository nature in the literature on the fluid dynamic principles necessary for understanding sedimentation in dilute and moderately concentrated dispersions [Happel and Brenner (1973); Batchelor (1972)]; however, there is as yet very little published work, on the theoretical side, on the colloidal interactions. In view of this, this review will be confined to a discussion of some elementary aspects of colloidal effects on settling of monodispersed suspensions.

The approach that is generally used for obtaining the settling velocity as a function of volume fraction is to expand the configurational average of the velocity in a virial series. As in the case of thermodynamic properties discussed in the previous Chapter, n-body interactions give rise to an n-th order term in the expansion. In practice, however, the expansion can be determined only up to the first-order term since higher-order hydrodynamic effects are barely understood. The colloidal interactions enter through the configurational averaging step since the interaction forces determine the local distribution of the particles. Since the effect of increased concentration of the particles is to effectively confine each to a bounded region, the flow field generated by the neighbors has a significant influence on the settling velocity of any given particle. Both enhancement and hindering can occur depending on whether the test particle experiences the downflow of the neighbor or the latter's backflow (which is caused by the container walls; see Figure 7.B.1). Reed

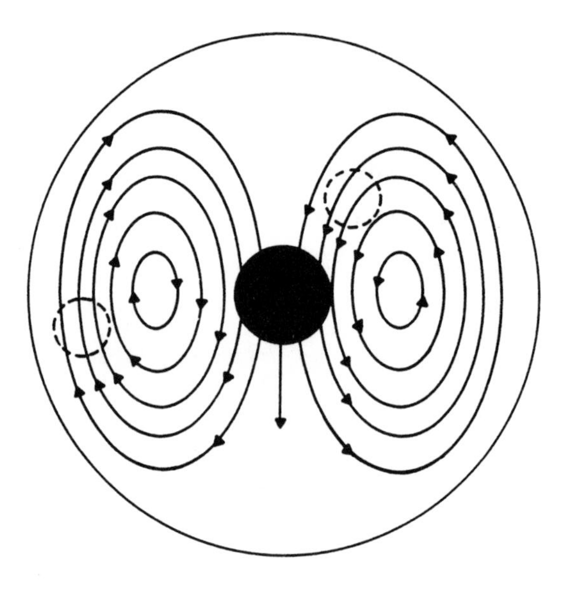

Figure 7.B.1 Fluid streamlines for particle motion in a bounded region [from Reed and Anderson (1980)].

A neighbor's velocity may be enhanced or hindered depending on its position in the flow.

and Anderson (1980) have evaluated the first-order coefficient in the virial expansion for monodispersed suspensions of charged spheres (e.g., globular proteins). The local distribution of particles has been assumed to be given by the 'dilute-gas' radial distribution function, and the configurational averaging of the settling velocity is then straightforward. An example of the extent of hindrance is given in Figure 7.B.2 as a function of the charge of the particles and the screening length. The colloidal interaction potential used for these results includes both a screened repulsion (of the Yukawa-type) and unretarded van der Waals attraction. An interesting observation that follows from the results shown in Figure 7.B.2 is that large inter-particle separations caused by repulsion (e.g., at low values of screening; see the Figure) actually lead to greater <u>hindrance</u> to settling since a typical particle finds itself, on the average, in its neighbor's backflow.

In the case of concentrated systems, the hydrodynamic interactions are assumed to be pairwise additive. The hydrodynamic interactions between the neighbors are preaveraged first in the absence of the test particle, and the test particle is then allowed to sample various locations within the suspension. Its interaction with the neighbors now becomes mathematically tractable in view of the assumption of pairwise additivity. (Notice that this ad hoc preaveraging technique is used because of the unavailability of expressions for hydrodynamic interactions for the general n-body problem with $n > 2$.) The values of the settling velocity computed by Reed and Anderson seem to be in very good agreement with some of the available experimental data for volume fractions up to 0.4. Reed and Anderson (1980) [and, prior to them, Goldstein and Zimm (1971)] have also suggested that

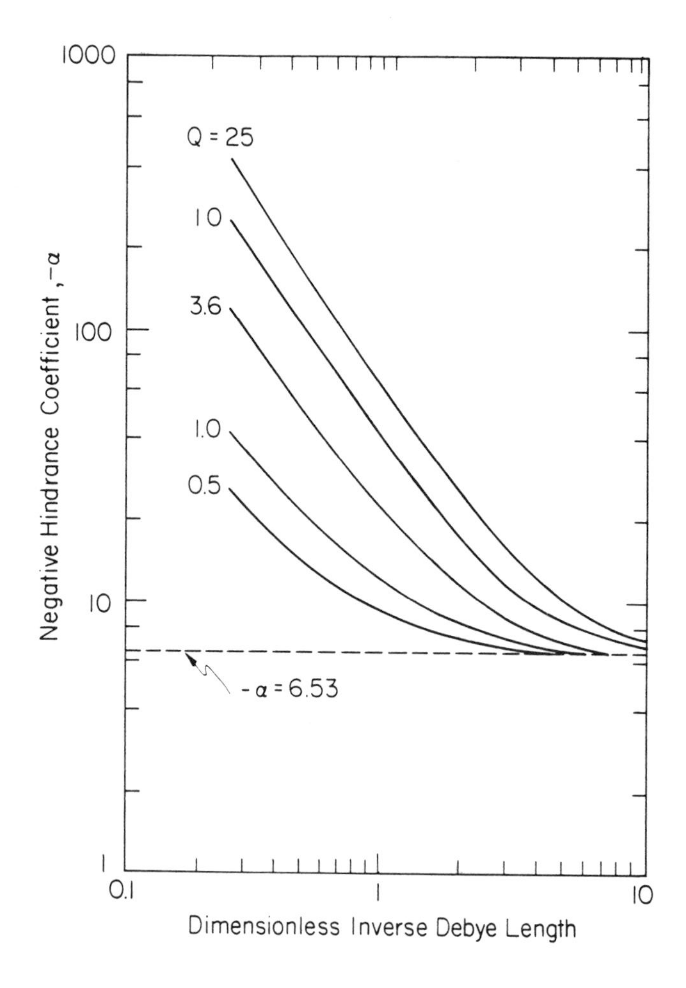

Figure 7.B.2 Sedimentation coefficient α as a function of screening length and charge of the particles [from Reed and Anderson (1980)].

The parameter Q is proportional to the square of the charges. The inverse screening length is made dimensionless with respect to the diameter of the particle.

the observed sedimentation rates can be used to extract information on colloidal interactions themselves (e.g., determination of Hamaker constants), but this, at present, is not established reliably.

The above discussion indicates that three physical phenomena influence sedimentation rates. These are (i) near-field hydrodynamic interactions, (ii) backflow of the fluid, and (iii) colloidal forces. The first two compete with each other, i.e., the former increases the rate while the latter retards sedimentation. The third does not exert any influence directly, but determines the local distribution of the particles, and hence the extent of influence of the other two. In view of this last comment, it is evident that a reasonably accurate representation of the radial distribution function may be necessary to assess correctly the influence of concentration on sedimentation. The paper by Reed and Anderson (1980) discussed above, and another paper by Dickinson (1980), have used a dilute-gas radial distribution function valid only at low concentrations. Glendinning and Russel (1982) circumvent this by using radial distribution functions obtained from the Percus-Yevick approximation of the integral equation approach common in liquid-state physics (see Chapter 6, Section B); however, they consider only hard-sphere suspensions. Nevertheless, their comparison of the effects of near-field hydrodynamics with those of backflow is instructive. The rapid decrease in the effect of backflow as the particles get closer to each other with increasing concentration is in sharp contrast to the slight increase in the effect of hydrodynamic interaction under the same conditions.

The use of radial distribution functions computed on the basis of colloidal interactions alone has been criticized by Batchelor (1982), who points out that the pair distribution function must be calculated from the solution of an appropriate Fokker-Planck equation which includes, in addition to Brownian diffusion and interaction forces, the influence of gravity as well. Batchelor notes that even when convection effects are masked by Brownian motion and the departure of the pair distribution function from the Boltzmann distribution is small, the perturbation due to gravity can lead to a non-negligible contribution to the mean velocity of the particles in the case of polydispersed suspensions. Solving the Fokker-Planck equation for the pair distribution function is, thus, an unavoidable part of investigations of non-equilibrium dynamical systems. Batchelor proceeds to derive analytical equations (up to first-order in concentration) for the effect of interaction between pairs of rigid spherical particles on the mean velocity of each species in a statistically homogeneous dilute polydisperse system. This derivation is a generalization of the earlier results for monodisperse systems [see Batchelor (1972) cited in the opening paragraph of this Section]. Results are presented for a range of special conditions such as large Peclet numbers (i.e., negligible effect of Brownian motion), small Peclet numbers and extreme values of the ratio of the radii of the two spheres. Numerical values based on these results are presented in Batchelor and Wen (1982). These include pair distribution functions for various values of Peclet numbers as well as sedimentation coefficients. The colloidal interaction potential considered is essentially an effective hard-sphere potential with a van der Waals tail. The effective diameter is determined by the screening of the repulsive potential. Batchelor and Wen offer the following conclusions, on the basis of their numerical results:

(i) A rough approximation to the sedimentation coefficient may be obtained by retaining only the direct contribution to the sedimentation coefficient due to gravity and ignoring those contributions due to interparticle forces and relative diffusion.

(ii) For accurate results, the pair distribution function needed for estimating the effect of gravity on sedimentation must include effects of interparticle forces and diffusion.

(iii) The effect of attractive forces on sedimentation can be calculated approximately by representing the radial distribution function in terms of an excess number of nearest neighbors due to attraction. [This was used in Batchelor (1972) for monodisperse systems.]

Additional details are available in the above papers.

C. Rheology of Interacting Dispersions

The obvious need to understand and quantify the flow behavior and properties of a wide class of suspensions because of their direct impact on process design and development has led, over the years, to an accumulation of a bewildering amount of both empirical and some fundamental information on the rheology of suspensions. Aside from their technological importance, dispersions of even simple structure often serve as sufficiently reasonable models of complex fluids and hence have attracted careful scrutiny. As emphasized by Turian (1982), implicit in the attribution of rheological properties to a given material is the assumption that the material can be treated as a continuum. Whether this assumption holds in the case of a manifestly discontinuous material such as a dispersion, and if not, what conditions need to be met for it to hold, depend on the spatial and temporal scales relevant to the particular physical situation under consideration.

The above comments are meant to draw attention to the logic that the research on rheology of dispersions has followed in recent years. Although, as mentioned earlier, a large body of empirical observations is available in the literature, this has been insufficient to shed light on the microscale origin of the complex macroscale behavior displayed by interacting dispersions. Attempts to improve this situation and to construct deductive theories have taken various forms. These include methods in which one uses detailed analyses of two-body interactions with pair distribution functions determined by the flow field [e.g., Batchelor and Green (1972a,b)] and, in the other extreme, methods that incorporate more complex interactions through cell models [Yaron and Gal-Or (1972); Herczynski and Pienkowska (1975)] or through near-contact dynamics [Frankel and Acrivos (1967)]. The former methods are valid strictly for dilute dispersions, while the cell models and the other methods are meant for more concentrated systems.

The microscale interactions in this case, as in the case of other transport properties discussed earlier, can again be classified as either hydrodynamic or colloidal. The references cited above address only the former, although the need to differentiate between 'dilute' and 'concentrated' dispersions, so that suitable lines of approach can be devised for each, is the same for both types of interactions. It is important to emphasize that this review will focus exclusively on situations in which 'colloidal' interactions dominate; hydrodynamic interactions, of course, cannot be ignored since they are always long-ranged and exert a significant influence except at extreme dilution. For those cases in which hydrodynamic interactions are the sole factor, excellent reviews and monographs are available [Bird, Warner and Evans (1971); Happel and Brenner (1973); Brenner (1974); Herczynski and Pienkowska (1980)]; in addition, a state-of-the-art review of this aspect of rheology of dispersions has been presented recently by Turian (1982). Heterogeneous systems in which the particles are larger than colloidal dimensions generally fall under this class. In fact, some investigators differentiate these systems from those in which colloidal forces are important by labeling the former as 'suspensions' and by reserving the term 'dispersions' for the latter [see Mewis and Spaull (1976)]; we do not use this distinction in this report, and the two terms are used interchangeably here.

This section is organized as follows. After a brief discussion which highlights the need to consider colloidal effects in the rheology of dispersions, a few major areas and phenomena that have attracted the attention of researchers in recent years are identified. Following this, a brief summary of the current status of understanding in each of those aspects is given. As in previous chapters, the research needs are summarized at the end of this Chapter.

C.1 Experimental Observations of Colloidal Effects

The influence of colloidal forces even at high ionic strengths can be seen in the experimental results of Krieger (1972) and Krieger and Eguiluz (1976) obtained for polystyrene latex particles at fairly high particle concentrations (volume fractions up to 0.5). The moderate shear thinning observed at high ionic strengths changes drastically as ionic strength decreases (i.e., the range of repulsion increases); the long-range repulsion increases the effective viscosity at low shear rates and eventually gives rise to yield stresses. These observations are based on experiments on aqueous dispersions. Willey and Macosko (1978) have presented some data on a nonaqueous system of polyvinyl chloride (PVC) plastisols. Their data indicate that in fluids that serve as θ-solvents for PVC molecules the dispersions exhibit slightly non-Newtonian viscosities. If the solvent is very good, the resulting stable dispersion (see Chapter 4, Section B) shows a significantly larger but finite viscosity. With poor solvents, on the other hand, instability causes flocculation and more viscous behavior (with eventual appearance of a yield stress) develops. Schoukens and Mewis (1978) have observed similar but stronger interactions with very low volume fractions of carbon blacks in mineral oils.

The above discussion is meant merely to illustrate the complex role of colloidal forces in rheology. A more extensive discussion of numerous experimental studies reported in the literature prior to 1976 is presented by Mewis and Spaull (1976). More recent accounts are available in Russel (1980) and Goodwin (1982). Goodwin's article also summarizes the experimental techniques that have been developed for studying rheological behavior of dispersions in a systematic manner. The types of general rheological behavior that can be expected in interacting dispersions are illustrated in Figure 7.C.1, and a summary of shear flow responses in viscometric flows is given in Table 7.C.1; see Hoffman (1983). Some observations on the rheology of concentrated dispersions from an industrial viewpoint and on its use for estimating the long-term stability of the dispersions are summarized in Tadros (1983). For our purpose here, the following summary of major observations to date is sufficient to highlight the need for fundamental research on rheology of dispersions.

(i) Stress levels in dispersions in which colloidal forces are important are much higher than can be expected from hydrodynamic interactions alone (see Figure 7.C.2). In addition to the magnitude of the shear stress, its change with shear rate is also important. Significant non-Newtonian behavior can also be expected at moderate or high particle concentrations (see Figure 7.C.3).

(ii) In addition to strong viscous behavior, plastic and elastic behavior can be expected in strongly-interacting dispersions. The existence of yield stresses in dispersions of low volume fractions is particularly significant since it cannot be explained by hydrodynamic and Brownian interactions alone. Presence of yield stresses can be explained in terms of thermodynamic arguments that appeal to interaction forces and entropic changes on shearing [see Hastings (1978)] and demonstrates the need to consider many-body interactions of colloidal nature. The ability to store mechanical energy reversibly during deformation (i.e., the existence of elastic behavior) also arises from solid-like behavior of some dispersions at low stresses and strains [see Mewis and Spaull (1976) for references to experimental data].

(iii) Some of the references cited above [Willey and Macosko (1978) on normal stresses and Schoukens and Mewis (1978) on non-linear viscoelastic behavior] and experimental data of van de Ven and Hunter (1979) on dynamic moduli have provided some quantitative information on elastic stresses in dispersions, and it appears that these stresses are more significant in dispersions in polymeric liquids [see also Goodwin (1975) and Russel (1980)].

(iv) Describing the above-mentioned rheological phenomena in terms of microscale colloidal interactions is extremely complex, but some

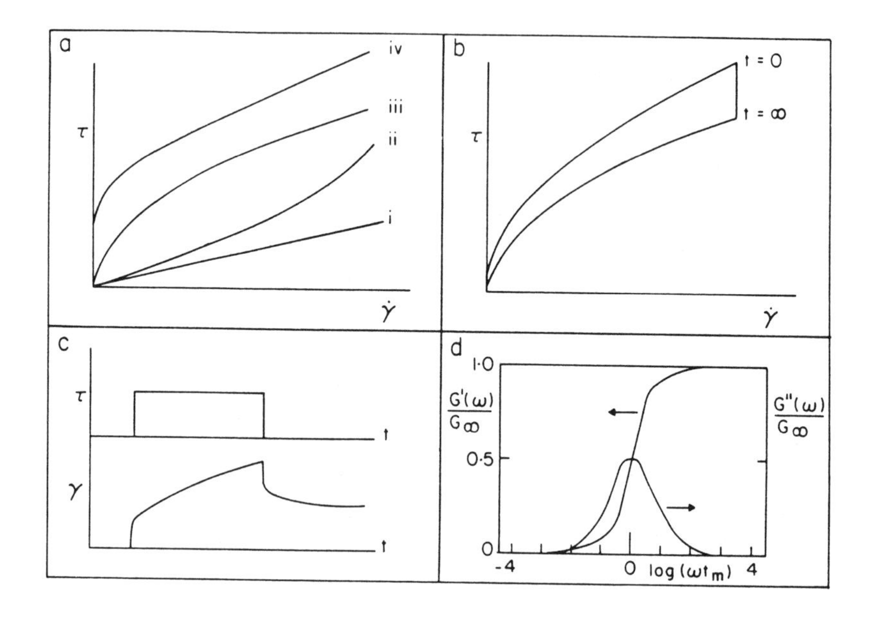

Figure 7.C.1 Typical rheological behavior of colloidal dispersions [see Goodwin (1982)].

a. (i) Newtonian; (ii) Shear thickening; (iii) Pseudoplastic; (iv) Plastic

b. Thixotropic

c. Creep response

d. Dynamic properties: relative storage modulus (G') and loss modulus (G'') as functions of dimensionless frequency. The Maxwell relaxation time is denoted by t_m.

Table 7.C.1
Steady Shear Flow Responses of Concentrated Suspensions in Viscometric Flows*

Dominant Forces	Shear Stress Level		
	Low	Intermediate	High
A. Hydrodynamic		1. Pseudo-Newtonian 2. Ordered layers (2-d ordering) 3. Spheres rotate "freely"	
B. Hydrodynamic/ Brownian	1. Low-shear Newtonian limit 2. Random order 3. Spheres rotate "freely"	1. Shear thinning 2. Transition in ordering 3. Spheres rotate "freely"	1. High-shear pseudo-Newtonian limit 2. Ordered layers (2-d ordering) 3. Spheres rotate "freely"
C. Hydrodynamic/ Attractive	1. Yield stress 2. Random order 3. Networks of flocced spheres (rotating?)	1. Shear thinning 2. Flocs ordered in layers? 3. Flocs tend to rotate as a unit 4. Transition to a smaller floc size	"
D. Hydrodynamic/ Repulsive	1. Yield stress 2. 3-dimensional order 3. Individual spheres (rotating?)	1. Very strong shear thinning 2. Transition to 2-dimensional ordering 3. Individual spheres rotate "freely"	"
E. Hydrodynamic Repulsive Attractive Frictional	1. Low-shear pseudo-Newtonian or shear thinning 2. Ordered layers (2-d order) 3. Sphere rotation hindered	1. Rheological dilatancy or discontinuous jump in viscosity 2. Random order 3. Groups of spheres rotation	"

*Adapted from Hoffman (1983).

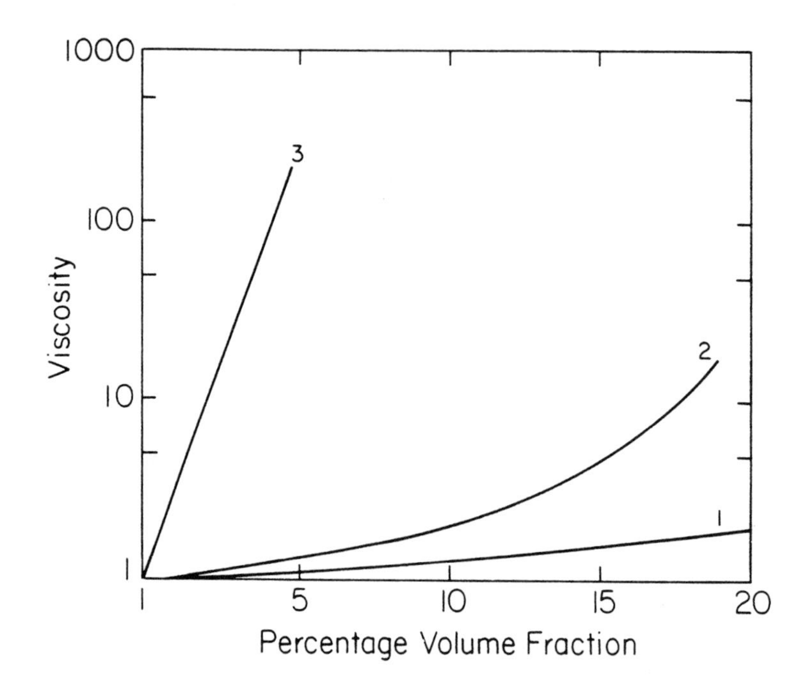

Figure 7.C.2 Stress levels in dispersions with and without colloidal interactions [from Mewis and Spaull (1976)].

Curve 1: Non-interacting dispersions

Curve 2: TiO_2 in stand oil

Curve 3: Carbon black in mineral oil

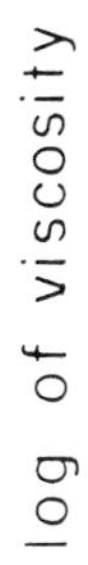

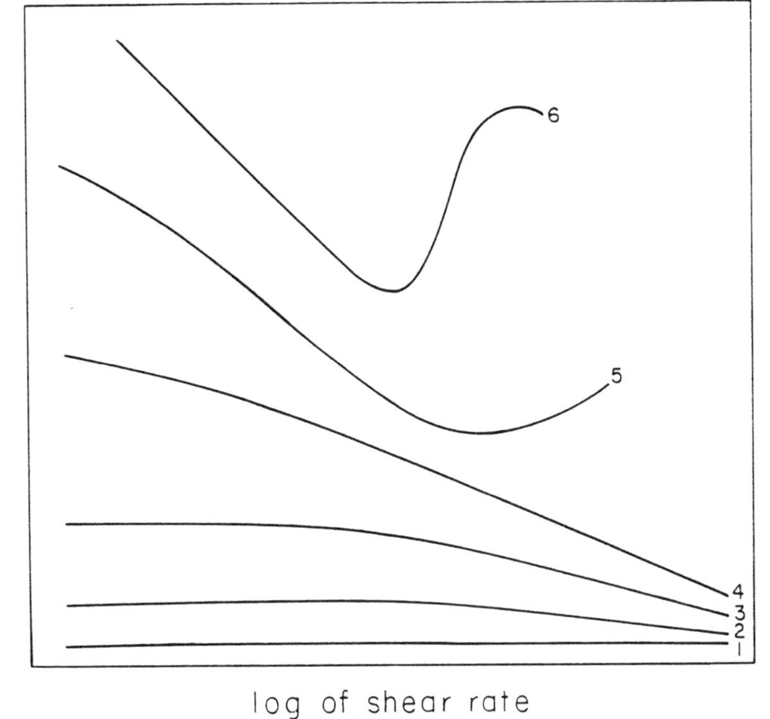

Figure 7.C.3 Shapes of the rheogram for various concentrations [from Mewis and Spaull (1976)].

Concentration increases from curves 1 to 6.

progress can be made by examining these under three special cases: (1) steric effects, (2) electrostatic effects, and (3) effects due to coagulation or flocculation.

We shall use the special cases mentioned in (iv) above to summarize the current research activities in this area.

C.2 Steric Effects on Rheology

Steric effects on stability are discussed in the Chapter on colloidal stability (Chapter 4, Section B) in this report. Most of the work reported in the literature on this aspect is concerned with static situations [see, for example, Mewis and Spaull (1976)], and there is very little work on the dynamic aspect in comparison to the status of electrostatic effects. Russel (1980) presents spatial and temporal scales relevant for estimating stability or lack of stability of dispersions; these are based on heuristic dimensional analysis and can be used to estimate the conditions under which the local equilibrium structure of the dispersions will be altered sufficiently to influence mechanical properties and to subsequently cause non-Newtonian behavior. In addition to providing a steric barrier to coagulation, adsorbed layers perturb the streamlines and hence affect the stress field. The usual method of studying the effects of steric layers on rheology has been to define an effective diameter for the particles or an effective volume fraction for the dispersion and then to use viscosity relations based on pure hydrodynamic interactions at the scaled volume fraction [Maron, Meadow and Krieger (1957); Woods and Krieger (1970); Krieger (1972)]. In fact, this technique is often used to estimate the thickness of the adsorbed layer from effective viscosity measurements, and the estimated thickness is then used for studying the steric stability of dispersions [Goodwin (1982)]. In this sense, this procedure illustrates a method of <u>using</u> rheological measurements for other purposes rather than predicting rheological behavior from information on steric stability. Some additional information on these can be found in Krieger (1972) and Goodwin (1975, 1982). Using equations based on hydrodynamic interactions at an effective volume fraction computed from adsorbed layer thickness is sufficient as long as the local structure is not affected by steric effects. If flocculation or doublet formation and chain formation occur, the rheological behavior changes considerably and this will be summarized in Section C.4 below. A special case, in which the conformational behavior of the adsorbed polyelectrolytes plays a role, is discussed in Section C.3 in the context of the tertiary electroviscous effect.

C.3 Rheology of Electrostatically Stabilized Dispersions

In contrast to the status in the area of sterically stabilized dispersions, the studies on electrostatically stabilized dispersions have carried the examination of colloidal forces on rheology considerably further in the direction of fundamental causes. This is not surprising since 'effective diameter' approaches are not sufficient in the latter case because of the long-range nature of colloidal forces and the electrohydrodynamics of interacting electrical double layers. Basically, three distinct electroviscous effects have been identified [see Conway and Dobry-Duclaux (1960)], and the mechanism assigned to each is considerably different from those of the others. The first of these, known as primary electroviscous effect, arises from the distortion of the diffuse part of the electrical double layer in the shear field. The secondary electroviscous effect, on the other hand, arises directly from inter-particle interactions which increase the effective excluded volume. Physically, the electrostatic repulsion between the particles provides another source of energy dissipation in the dispersion and therefore causes an increase in effective viscosity. The tertiary electroviscous effect is due to the expansion or contraction of particles (containing polyelectrolytes on the surface) with changes in chemical make-up of the suspending medium. For sterically stabilized dispersions, this problem leads to the determination of conformation behavior of the stabilizer molecules and its effect on effective volume fraction (which is needed in the type of approach described in Section C.2 above). This section presents a brief overview of the progress in understanding the electroviscous phenomena.

Following the pioneering work of Smoluchowski (1916) and Krasny-Ergen (1936), many investigators have studied the role of primary electroviscous effect in determining the viscosity for thick electrical double layers [Booth (1950)] and for curved double layers [Finkelstein and Cursin (1942)]. Booth's results apply to dilute dispersions at low shear rates and zeta potentials, and the Krasny-Ergen relation is restricted to thin double layers. All these lead to first-order corrections (in terms of volume fraction) to the viscosity as functions of the zeta potential, radius of the particle and the relative permittivity and specific conductivity of the continuous phase. A second-order correction is also available in the literature [Whitehead (1969)], and these are summarized in Goodwin (1975) along with some discussion of experimental tests of the resulting expressions.

The Krasny-Ergen result has been extended recently to stronger flows by Russel (1978, 1980) and first and second normal stress differences have been computed. Large distortions of the double layers in both simple shear and extensional flows have been treated by Lever (1979). These results lead to the following observations:

(i) At large shear rates, the primary electroviscous effect leads to shear thinning and normal stresses. Large distortions can also lead to elastic properties.

(ii) Experimental tests reveal that at least first-order correction to the viscosity can be estimated using the above theories. However, the primary electroviscous effect is usually small when compared to the others and it is doubtful if the currently available instrumental techniques are sufficient to test the predicted normal stresses.

The literature is even sparser in the case of secondary electroviscous effect. As mentioned earlier, this effect is due to direct inter-particle interactions arising from colloidal forces, which contribute to additional energy dissipation. This has been recognized in the literature [see, for example, Conway and Dobry-Duclaux (1960); Chan and Goring (1966); Chan, Blachford and Goring (1966); Blachford, Chan and Goring (1969)]; the last three references present one of the most successful treatments to date on this effect. This treatment is based on balancing the hydrodynamic forces on two particles (during collision and rotation) against forces due to overlapping double layers. The particles are assumed to interact at constant surface potentials and the screening of repulsion is taken to be low (i.e., thick double layers). Since this treatment considers the drag on the particles to be given by Stokes' law, the final result can only give a lower bound on the secondary electroviscous effect (i.e., the results will underestimate the severity of the effect). The experimental data of Stone-Masui and Watillon (1968, 1970) seem to fall below the viscosity predicted by the above-mentioned theory, but Russel (1978, 1980) reports better agreement with these data for the coefficient of the second-order term in the expansion of viscosity (in terms of volume fraction) calculated by a different method. His analysis assumes a dilute-gas type radial distribution for the particles (i.e., low particle concentration) and, because of the large average separation between the particles, further assumes that the van der Waals attraction is negligible; that is, the particles interact through a screened repulsive potential. The large inter-particle separations also permit one to neglect the hydrodynamic interactions, and the simplified differential equation for the distribution of the particles in the low-shear limit is solved by matched asymptotic expansion. (The high-shear limit is, however, dominated by hydrodynamic interactions.) The results discussed above are strictly valid only at low volume fractions and, in addition, only if the dispersions do not pass into ordered structures at equilibrium. The latter case, generally called 'concentrated' systems, is discussed following a brief comment concerning the tertiary electroviscous effect.

The major difficulty in the case of the <u>tertiary electroviscous effect</u> is the determination of conformation of the polyelectrolyte chains on the surfaces of the particles. The extent of penetration of the moving fluid in the adsorbed layer and its effect on the boundary conditions for the flow field are complex and still await examination.

The discussion so far applies to fairly dilute, moderately-interacting dispersions with liquid-like or gas-like structures. The <u>crystal-like structures</u> that form in colloidal materials were discussed in Chapter 6 of this report. Experimental investigations of such crystal-like structures and liquid-like structures at high volume fractions have employed oscillatory viscometers [Russel (1980)] for less elastic suspensions and propagation velocity of shear waves [Goodwin and Khidher (1976)] for others. These lead to storage and loss moduli, in the case of the former, and to rigidity modulus, in the latter. Other measurements, cited in the previous Chapter of this report, include osmotic pressure, osmotic compressibility [Homola and Robertson (1976)] and elasticity [through compression of the crystal estimated from diffraction measurements; Crandall and Williams (1977)]. Theoretical studies of these require reliable methods for accounting for many-body interactions. Three techniques have appeared in the literature, namely, Monte Carlo simulations [van Megen, Snook and Watts (1980)], 'static structure models' in which the dispersion is treated as a static array [but without assuming long-range order; Goodwin and Smith (1974); Goodwin and Khidher (1976); Goodwin (1982)], and a mean field approach [Russel (1980)]. The first two employ pair-wise additivity of interaction potential. While the first is a computer simulation method and does not lead to closed-form expressions, the second (the static structure model) offers the convenience of a simple algebraic equation for the shear modulus in terms of the volume fraction, particle radius and pair potential of interaction. The value computed on the basis of the latter agrees well with the more accurate value based on computer experiments. The mean field approach avoids the assumption of pair additivity by treating the many-body interactions in terms of interaction of a charged sphere imbedded in a continuum having the mechanical and electrical properties of the bulk suspension. Classical conservation equations can now be used to obtain osmotic pressures, shear modulus, dynamic viscosity, dielectric constant, and conductivity. Russel (1980) reports good agreement of these calculations with measured osmotic pressures. Recently, Goodwin (1982) has also reported attempts to develop a model for viscosity at low shear stresses.

The cell model of Frankel and Acrivos (1967) cited in the beginning of Section C, which was originally developed for concentrated dispersions interacting only hydrodynamically, can be extended to include the effects of colloidal interactions. Such an attempt has been made recently by Tanaka and White (1980), who first extend the cell model to dispersions in power-law fluids and then to dispersions interacting through van der Waals and electrostatic forces. Equations for effective viscosity have been developed at both low shear rates and high shear rates by computing the total energy dissipation due to hydrodynamic and colloidal effects. (The carrier fluid has been assumed to be Newtonian at low shear rates and to be of the power-law type at high shear rates.) Yield stresses have also been computed and have been compared with earlier experimental results. At low shear, the flow behavior is controlled by colloidal forces, and, at high shear, hydrodynamic interactions and effects of shape of the particles take control, as would be expected intuitively.

In addition to the above classical approach, a few statistical mechanical theories of rheology of interacting dispersions available in the literature (other than those mentioned earlier) also deserve attention. Evans and Watts (1980) have used a non-equilibrium molecular dynamics method reported by Evans (1979) to simulate a dispersion under homogeneous shear. Using a combination of statistical mechanics and rheological framework, they derive an equation for shear viscosity for interacting particles. This derivation is analogous to that of the equation for osmotic pressure in terms of the radial distribution function, except that, in this case, the pressure is replaced by the stress tensor and the radial distribution function is dependent on the rate of strain. Evans and Watts report the viscosity as a function of shear rates and show that the dispersion exhibits shear dilatancy in conjunction with shear thinning. The simulated effects occur at high shear rates and do not include hydrodynamic interactions; consequently the relevance of the observations of

Evans and Watts to real measurements is not clear [see, also, Dickinson (1983)]. Another statistical approach, based on the assumption that the flow of the dispersion can be treated as a cooperative phenomenon through the Ising model, has been proposed by Bohlin (1980). Bohlin's theory is an extension of the earlier work of Bohlin and Kubat (1976) on stress relaxation in solids. Bohlin assumes that the dispersion can be divided into a collection of flow units, each of which can exist in either a relaxed or a stressed state. The development then proceeds from a kinetic equation for the states of the flow units to the solution for the average state of the system through a series of approximations. The relative stress thus obtained depends on three parameters, namely, the coordination number, the difference in energy between the stressed and the relaxed state and the 'energy of interaction' between the neighboring elements. Since, except for the coordination number, there is no clear way of defining the other two parameters in terms of colloidal interactions, Bohlin suggests a graphical procedure for determining these parameters from experimental relaxation data. Nevertheless, Bohlin's theory offers a novel, and potentially useful, approach to the study of flow of dispersions.

Finally, an investigation which aims for a direct examination of structural changes introduced by shear has been initiated by Ackerson and Clark (1981, 1983) in an attempt to use colloids as model many-body systems. These studies, using dynamic light scattering for non-intrusive and direct examination of local structure, seek to employ model colloids and colloidal crystals for eventually studying the behavior of <u>pure atomic systems</u> under extreme conditions (e.g., flow under geological conditions); these are discussed further in Chapter 6, Sections A.3 and B.5.

C.4 Effects of Flocculation on Rheology

The previous discussions on the complexity of rheological behavior in dispersions of reasonably well-defined structures certainly give an indication of the task one faces in the case of suspensions of coagulated and flocculated materials. In addition to factors such as shape, size and strength of the flocs, a point of major interest here, particularly at low shear rates, is how the dynamics of flocculation itself is affected by the flow, since this certainly has an influence on the macroscopically observed behavior. For example, the presence of secondary electroviscous effect implies that there are no longer closed trajectories when two particles interact and, in addition, the paths of approach and recession differ from each other (see Figure 7.C.4). These differences from the case of uncharged dispersions have a direct influence on coagulation rates (see Figure 7.C.5), as well as on the subsequent rheological consequences. This (highly simplified) example draws attention to the interplay between stability considerations and rheology in the case of charged particles; this is precisely the reason why these are almost always discussed together in the literature on rheology of suspensions of flocs.

A large amount of experimental information on these is available in the literature. Mewis and Spaull (1976) discuss the empirical and some theoretical data available on both flocculation kinetics and rheology and cite over fifty references on these alone. Nevertheless, the current level of understanding, particularly on low shear flows, is much too vague and nebulous to provide information on the fundamental interactions. Perhaps the dynamic light scattering techniques of Ackerson and Clark (1981, 1983) mentioned in the previous section may be extended to provide both structural and

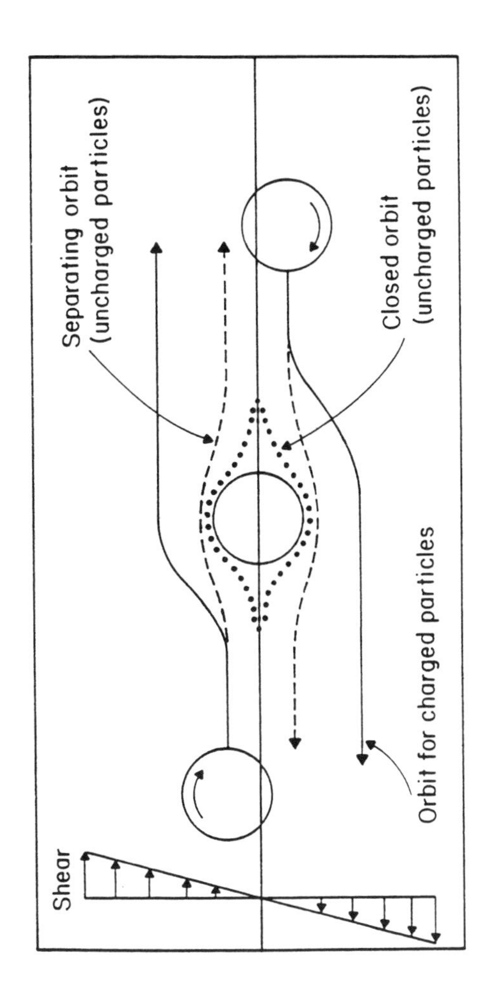

Figure 7.C.4 Trajectories for charged and uncharged particles in shear flow.

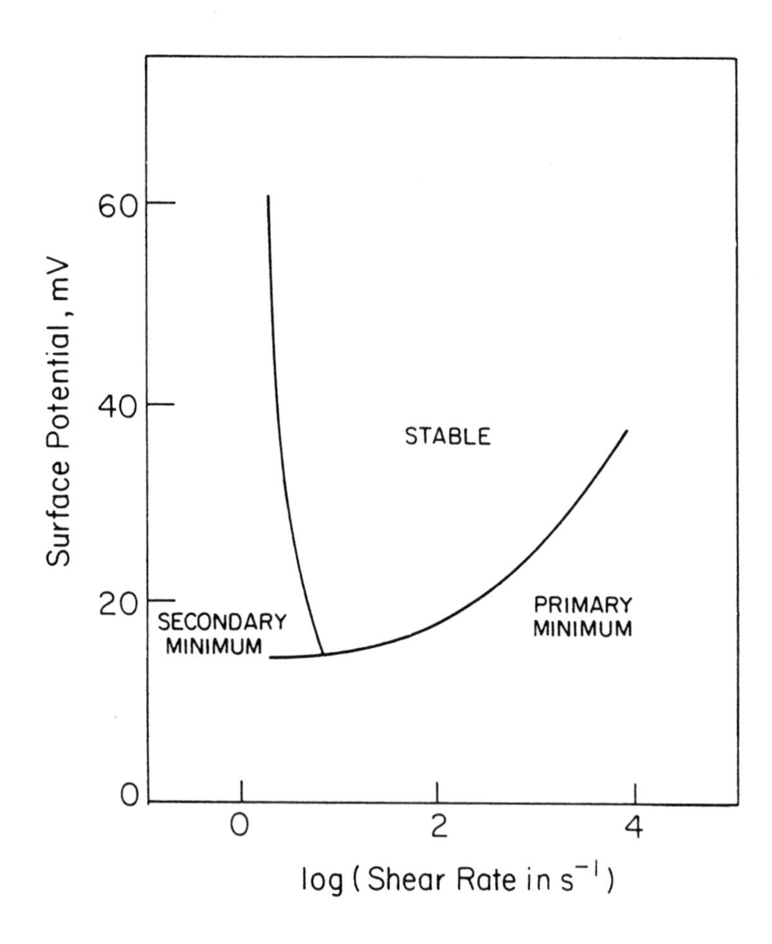

Figure 7.C.5 Effect of shear rate and surface potential of the particles on stability [see Russel (1980)].

As shear rate increases, dispersions flocculated in the secondary minimum may become stable first and may subsequently coagulate in the primary minimum. If the primary minimum is shallow, additional increases in the shear rate may make the dispersion stable again.

rheological information in these cases. It is possible that other techniques such as the dielectric characterization of local structure attempted by Helsen et al. (1978) might be of value as well.

The investigations on the high-shear limit, however, seem to be leading to tractable results. Of these, the following deserve attention:

(1) effects of flow and colloidal forces on the rate of coagulation and flocculation in the high-shear limit, and

(2) the rheological consequences.

The first of these was the subject of discussion in Chapter 4, Section C, and the second will be taken up here.

The general rheological features of flocculated dispersions have been noted for a long time; for instance, the general viscoelastic behavior of shear stress as a function of shear rate in flocculated suspensions is sketched in Figure 7.C.6 [see Michaels and Bolger (1962); Friend and Hunter (1971); Firth and Hunter (1976a,b,c)]. Other representations are also available in the literature [e.g., Jeffrey and Acrivos (1976)], but the analyses of Gillespie (1960) and Russel (1980) seem to provide some justification for the qualitative picture shown in Figure 7.C.6. Russel's analysis considers hard spheres and temporary doublets in a model suspension in which the hard spheres are subjected to convection and short-range attraction while the doublets are assumed to merely rotate with the flow. The balance between viscous and interparticle forces on the doublets determines the critical shear rate shown in the figure. The shear stress has been determined up to second order in volume fraction and is in qualitative agreement with the behavior shown in the figure. The final form of the stress is close to that derived by Gillespie (in a much less rigorous, but more intuitive fashion). However, the

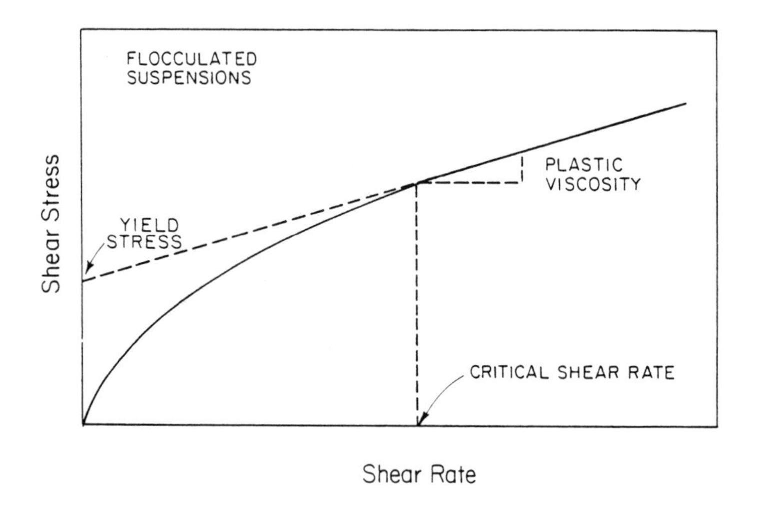

Figure 7.C.6 Typical shear stress-shear rate curves for flocculated suspensions.

experimental results of Firth and Hunter (1976a,b,c) cited earlier have already demonstrated that the quantitative agreement between experimental results based on model colloids and Gillespie's equation is quite poor. The substantially higher plastic viscosities observed experimentally indicate that the interacting units cannot be assumed to be individual particles. This has led to semi-empirical corrections for floc volumes and interactions, with only partial success. van de Ven and Hunter (1977) have tried to improve this by postulating 'elastic floc' models which account for deformation and attrition of the flocs, but these improvements require semi-empirical estimations of intermediate parameters. Nevertheless, the progress in this area is encouraging and points to the need for well-defined experiments and theoretical analyses of individual components of the problems, such as interactions between flocs and mechanics and structure of flocs.

Research Needs

Empirical information available in the literature on transport properties of concentrated dispersions is rather extensive, but systematic studies on the effects of colloidal interactions are of only recent origin. As mentioned in the overview section of this Chapter, the combination of hydrodynamic interactions with colloidal effects complicates the analysis in these cases. However, experimental and theoretical techniques for studying these effects separately (to a limited extent) and collectively are becoming available. In the following, we list some major outstanding problems that require further attention.

- Approximations for hydrodynamic interactions (such as pair-additivity, screened interactions, and effective-mobilities) need further study. A combination of computer experiments and experimental measurements of diffusion coefficients in neutral, hard-sphere suspensions is necessary to accomplish this. Development of tracer-particle techniques (such as the one discussed in Section A) is expected to improve the experimental studies of hydrodynamic interactions.

- Brownian dynamics experiments are possible in principle, but not practicable because of time-consuming computations. Simulations with a few particles are possible (and are being carried out) and are expected to be useful.

- More experiments on sedimentation rates of particles in model colloids are needed. These will complement small-angle radiation scattering measurements of mutual diffusion coefficients.

- Additional theoretical work on computation of sedimentation coefficients, for realistic pair potentials, is needed. Currently available results are restricted to modifications of hard-sphere potentials. Computations available for more realistic potentials do not account for the effects of flow field on local configurations.

- Experimental work using multiprobe techniques and other related techniques is needed and is discussed in Chapter 6.

- Viscous behavior of liquid-like dispersions has not received enough attention. This is considerably more complicated (certainly from the theoretical side) than the corresponding situation in the case of ordered structures. Measurements of wave rigidity modulus and measurements of the viscous part of creep compliance will be useful for understanding interaction energy profiles in concentrated dispersions [see, for example, Goodwin (1982) cited in Section C].

- Viscoelastic materials produce normal stresses. Although the measurements of normal forces have been carried out extensively in polymer rheology, they have not been investigated in colloidal dispersions.

- Systematic measurements of yield stresses in model colloidal crystals are not available, but can be performed with available techniques [see Section C.1].

- Studies of structure of flocs and of their mechanical behavior are needed.

- Transport properties of dispersions in the presence of other objects (e.g., in membranes) have not been discussed in this Report, but are important.

- There is very little work of the types discussed above for dispersions in non-aqueous systems. In sterically-stabilized dispersions, the tertiary electroviscous effect needs more attention.

REFERENCES: SECTION A. DIFFUSION COEFFICIENTS

Ackerson, B. J., J. Chem. Phys. 64, 242 (1976).

Altenberger, A. R., Chem. Phys. 15, 269 (1976).

Beenakker, C. W. J., Physica A, in press (1984).

Beenakker, C. W. J. and Mazur, P., Phys. Lett. A 91, 290 (1982).

Beenakker, C. W. J. and Mazur, P., Phys. Lett. A 98, 22 (1983).

Beenakker, C. W. J. and Mazur, P., Physica A, in press (1984).

Corti, M. and Degiorgio, V., pp. 111-124 in Degiorgio, Corti and Giglio (1980).

Dahneke, B. E., (ed.), Measurement of Suspended Particles by Quasi-Elastic Light Scattering, Wiley, New York, NY, 1983.

Degiorgio, V., Corti, M. and Giglio, M., Light Scattering in Liquids and Macromolecular Solutions, Plenum, New York, NY, 1980.

Doherty, P. and Benedek, G. B., J. Chem. Phys. 61, 5426 (1974).

Felderhof, B. V., J. Phys. A 11, 929 (1978).

Glendinning, A. B. and Russel, W. B., J. Colloid Interface Sci. 89, 124 (1982).

Hanna, S., Hess, W. and Klein, R., Physica (Utrecht) A 111, 181 (1982).

Hess, W., pp. 31-50 in Degiorgio, Corti and Giglio (1980).

Hess, W. and Klein, R., Adv. Phys. 32, 173 (1983).

Kitchen, R. G., Preston, B. N. and Wells, J. D., J. Polym. Sci. Symp. 55, 39 (1976).

Kops-Werkhoven, M. M. and Fijnaut, H. M., J. Chem. Phys. 77, 2242 (1982).

Kops-Werkhoven, M. M., Pathmamanoharan, C., Vrij, A. and Fijnaut, H. M., J. Chem. Phys. 77, 5913 (1982).

Mazur, P. and van Saarloos, W., Physica (Utrecht) A 115, 21 (1982).

Nicoli, D. F. and Dorshow, R. B., pp. 501-527 in Dahneke (1983).

Pecora, R. (ed.), Dynamic Light Scattering and Velocimetry: Applications of Photon Correlation Spectroscopy, Plenum, New York, NY, 1982.

Phillies, G. D. J., J. Chem. Phys. 60, 976 (1974).

Pusey, P. N., J. Phys. A: Math. Gen. 11, 119 (1978).

Pusey, P. N., Phil. Trans. Roy. Soc. London A 293, 429 (1979).

Pusey, P. N., Fijnaut, H. M. and Vrij, A., J. Chem. Phys. 77, 4270 (1982).

Pusey, P. N. and Tough, R. J. A., in Pecora (1982).

Schor, R. and Serrallach, E. N., J. Chem. Phys. 70, 3012 (1979).

Snook, I., van Megen, W. and Tough, R. J. A., J. Chem. Phys. 78, 5825 (1983).

Stephen, M. J., J. Chem. Phys. 55, 3878 (1971).

Stigter, D. and Mysels, K. J., J. Phys. Chem. 59, 45 (1955).

Stigter, D., Williams, R. J. and Mysels, K. J., J. Phys. Chem. 59, 330 (1955).

van Megen, W., Snook, I. and Pusey, P. N., J. Chem. Phys. 78, 931 (1983).

van Saarloos, W. and Mazur, P., Physica (Utrecht) A 120, 77 (1983).

Venkatesan, M., Hirtzel, C. S. and Rajagopalan, R., Effect of Many-Body Interactions on the Self-diffusion Coefficients of Colloidal Particles in Concentrated Dispersions, Paper presented at the 57th ACS Colloid and Surface Science Symposium, Toronto, Canada, June 1983.

Venkatesan, M., Hirtzel, C. S. and Rajagopalan, R., Long-time Self-diffusion Coefficients in Strongly-interacting Dispersions, Paper presented at the 58th ACS Colloid and Surface Science Symposium, Pittsburgh, PA, June 1984.

Verlet, L. and Weis, J. J., Phys. Rev. A 5, 939 (1972).

Wiersema, Y., Wiersema, P. H. and Overbeek, J. Th. G., Proc. K. Ned. Akad. Wet. Ser. B 72, 17 (1969a).

Wiersema, Y., Wiersema, P. H. and Overbeek, J. Th. G., Proc. K. Ned. Akad. Wet. Ser. B 72, 29 (1969b).

REFERENCES: SECTION B. SEDIMENTATION IN INTERACTING DISPERSIONS

Batchelor, G. K., J. Fluid Mech. 52, 245 (1972).

Batchelor, G. K., J. Fluid Mech. 119, 379 (1982).

Batchelor, G. K. and Wen, C-S., J. Fluid Mech. 124, 495 (1982).

Dickinson, E., J. Colloid Interface Sci. 73, 578 (1980).

Glendinning, A. B. and Russel, W. B., J. Colloid Interface Sci. 89, 124 (1982).

Goldstein, B. and Zimm, B. H., J. Chem. Phys. 54, 4408 (1971).

Happel, J. and Brenner, H., Low Reynolds Number Hydrodynamics, Martinus Nijhoff, The Hague, The Netherlands, 1973.

Reed, C. C. and Anderson, J. L., AIChE J. 26, 816 (1980).

REFERENCES: SECTION C. RHEOLOGY OF INTERACTING DISPERSIONS

Ackerson, B. J. and Clark, N. A., Phys. Rev. Lett. 46, 123 (1981).

Ackerson, B. J. and Clark, N. A., Physica A 118, 221 (1983).

Batchelor, G. K. and Green, J. T., J. Fluid Mech. 56, 375 (1972a).

Batchelor, G. K. and Green, J. T., J. Fluid Mech. 56, 401 (1972b).

Bird, R. B., Warner, H. R. and Evans, D. C., Adv. Polymer Sci. 8, 1 (1971).

Blachford, J., Chan, F. S. and Goring, D. A. I., J. Phys. Chem. 73, 1062 (1969).

Bohlin, L., J. Colloid Interface Sci. 74, 423 (1980).

Bohlin, L. and Kubát, J., Solid State Commun. 20, 211 (1976).

Booth, F., Proc. Roy. Soc. A 203, 533 (1950).

Brenner, H., Int. J. Multiphase Flow 1, 195 (1974).

Chan, F. S., Blachford, J. and Goring, D. A. I., J. Colloid Interface Sci. 22, 378 (1966).

Chan, F. S. and Goring, D. A. I., J. Colloid Interface Sci. 22, 371 (1966).

Conway, B. E. and Dobry-Duclaux, A., pp. 83-120 in Eirich (1960).

Crandall, R. S. and Williams, R., Science 198, 293 (1977).

Dickinson, E., pp. 150-179 in Everett (1983).

Eirich, F. (ed.), Rheology, Theory and Applications, Vol. 3, Academic Press, New York, NY, 1960.

Evans, D. J., Mol. Phys. 37, 1745 (1979).

Evans, D. J. and Watts, R. O., Chem. Phys. 48, 321 (1980).

Everett, D. H. (ed.), Specialist Periodical Reports, Colloid Science, Vol. 2, The Chem. Soc., London, U. K., 1975.

Everett, D. H. (ed.), Specialist Periodical Reports, Colloid Science, Vol. 4, The Roy. Soc. Chem., London, U. K., 1983.

Finkelstein, B. N. and Cursin, M. P., Acta Physicochim. URSS 17, 1 (1942).

Firth, B. A. and Hunter, R. J., J. Colloid Interface Sci. 57, 248 (1976a).

Firth, B. A. and Hunter, R. J., J. Colloid Interface Sci. 57, 257 (1976b).

Firth, B. A. and Hunter, R. J., J. Colloid Interface Sci. 57, 266 (1976c).

Frankel, N. A. and Acrivos, A., Chem. Eng. Sci. 22, 847 (1967).

Friend, J. P. and Hunter, R. J., J. Colloid Interface Sci. 37, 548 (1971).

Gillespie, T., J. Colloid Sci. 15, 219 (1960).

Goodwin, J. W., pp. 246-293 in Everett (1975).

Goodwin, J. W., pp. 165-195 in Colloidal Dispersions, Goodwin, J. W. (ed.), The Roy. Soc. Chem., London, U. K., 1982.

Goodwin, J. W., and Khidher, A. M., pp. 529-547 in Kerker (1976).

Goodwin, J. W. and Smith, R. W., Faraday Discuss. Chem. Soc. 57, 126 (1974).

Happel, J. and Brenner, H., Low Reynolds Number Hydrodynamics, Martinus Nijhoff, The Hague, The Netherlands, 1973.

Hastings, R., Phys. Lett. A 67, 316 (1978).

Helsen, J. A., Govaerts, R., Schoukens, G., de Graeuwe, J. and Mewis, J., J. Phys. E 11, 139 (1978).

Herczynski, R. and Pienkowska, I., Arch. Mech. Eng. (Warsaw) 27, 201 (1975).

Herczynski, R. and Piensowska, I., Ann. Rev. Fluid Mech. 12, 237 (1980).

Hoffman, R. L., pp. 578-588 in Poehlein, Ottewill and Goodwin (1983).

Homola, A. and Robertson, A. A., J. Colloid Interface Sci. 54, 286 (1976).

Jeffrey, D. J., and Acrivos, A., AIChE J. 22, 417 (1976).

Kerker, M. (ed.), Colloid and Interface Science, Vol. IV, Academic Press, New York, NY, 1976.

Krasny-Ergen, B., Zolloid Z. 74, 172 (1936).

Krieger, I. M., Adv. Colloid Interface Sci. 3, 111 (1972).

Krieger, I. M. and Eguiluz, M., Trans. Soc. Rheol. 20, 29 (1976).

Lever, D. A., J. Fluid Mech. 92, 421 (1979).

Maron, S. H., Madow, B. P. and Krieger, I. M., J. Colloid Sci. 6, 584 (1957).

Mewis, J. and Spaull, A. J. B., Adv. Colloid Interface Sci. 6, 173 (1976).

Michaels, A. S. and Bolger, J. C., Indus. Eng. Chem. Fundam. 1, 153 (1962).

Poehlein, G. W., Ottewill, R. H. and Goodwin, J. W. (eds.), Science and Technology of Polymer Colloids: Characterization, Stabilization and Application Properties, Vol. II, Martinus Nijhoff, The Hague, The Netherlands, 1983.

Russel, W. B., J. Fluid Mech. 85, 673 (1978).

Russel, W. B., J. Rheol. 24, 287 (1980).

Schoukens, G. and Mewis, J., J. Rheol. 22, 381 (1978).

Smoluchowski, M., Zolloid Z. 18, 194 (1916).

Stone-Masui, J. and Watillon, A., J. Colloid Interface Sci. 28, 187 (1968).

Stone-Masui, J. and Watillon, A., J. Colloid Interface Sci. 34, 327 (1970).

Tadros, Th. F., pp. 531-551 in Poehlein, Ottewill and Goodwin (1983).

Tanaka, H. and White, J. L., J. Non-Newtonian Fluid Mech. 7, 333 (1980).

Turian, R. M., Slurry and Suspension Transport, Report No. NSF/OIR-82001, National Science Foundation, Washington, D. C., 1982.

van de Ven, T. G. M. and Hunter, R. J., Rheol. Acta 16, 534 (1977).

van de Ven, T. G. M. and Hunter, R. J., J. Colloid Interface Sci. 68, 135 (1979).

van Megen, W. J., Snook, I. K. and Watts, R. O., J. Colloid Interface Sci. 77, 131 (1980).

Whitehead, J. R., J. Colloid Interface Sci. 30, 424 (1969).

Willey, S. J. and Macosko, C. W., J. Rheol. 22, 525 (1978).

Woods, M. E. and Krieger, I. M., J. Colloid Interface Sci. 34, 91 (1970).

Yaron, I. and Gal-Or, B., Rheol. Acta 11, 241 (1972).

SUMMARY

8

SUMMARY

CHAPTER 8

SUMMARY

As mentioned in Chapter 1, the scope of this Report is restricted to dispersions of rather well-defined particles, and even there to certain classes of problems. This decision is not based on any perceived importance of the material covered here over what has been left out. The selection of the topics presented here has been dictated by our own interests; however, we believe that these topics address a large body of current research in colloid science. In addition, the topics discussed illustrate a few basic points we have set out to emphasize, namely, the need for fundamental research in colloidal interactions, the need for interdisciplinary interactions and the possibility of cross-disciplinary impact of such interactions.

The discussions on intermolecular, electrokinetic and other forces presented in this Report are also relevant to topics that are not covered here (for example, adhesion and wetting). Further, although we did not discuss interactions in micellar systems and microemulsions explicitly, the materials presented in Chapters 6 and 7 are relevant to such dispersions as well. In fact, some such applications have been mentioned in passing in these Chapters, and the references cited therein contain additional information.

There are many areas of colloid science not covered here that need consolidation and which may benefit from reviews such as this one. A partial list of these areas is presented below.

- The thermodynamics of association colloids is an area that has grown rapidly in recent years. Some excellent reviews are available on this topic, and many novel experimental techniques are being used for studying the thermodynamics (as well as the kinetics) of formation of association colloids. The same can be said of microemulsions. The phenomena in this class have a direct impact on many biological and biomedical processes and must be studied in combination with the topics discussed in this Report.

- Thin film phenomena, which include foams and foam stability and contact angle phenomena, also merit separate study.

- Interfacial phenomena such as stability of fluid/fluid interfaces, properties of interfaces, and transport phenomena at interfaces, comprise another major area that should be treated separately.

- Aerocolloidal systems constitute another class of practically-important and fundamentally-challenging problems that deserve review.